Annals of Mathematics Studies

Number 90

INFINITE LOOP SPACES

BY

J. F. ADAMS

Hermann Weyl Lectures
The Institute for Advanced Study

PRINCETON UNIVERSITY PRESS
AND
UNIVERSITY OF TOKYO PRESS

PRINCETON, NEW JERSEY
1978

Published in Japan exclusively by
University of Tokyo Press;
In other parts of the world by
Princeton University Press

Printed in the United States of America
by Princeton University Press, Princeton, New Jersey

Library of Congress Cataloging in Publication data will
be found on the last printed page of this book

HERMANN WEYL LECTURES

The Hermann Weyl Lectures are organized and sponsored by the School of Mathematics of the Institute for Advanced Study. Their aim is to provide broad surveys of various topics in mathematics, accessible to nonspecialists, to be published eventually in the Annals of Mathematics Studies.

The present monograph is the third in the series. It is an outgrowth of the sixth set of Hermann Weyl Lectures, which consisted of six lectures given by Professor J. F. Adams at the Institute for Advanced Study on March 18, 19, 24, 26, April 1, 2, 1975.

ARMAND BOREL

JOHN W. MILNOR

PREFACE

This book derives from a series of Hermann Weyl Lectures which I gave at the Institute for Advanced Study, Princeton, in the spring of 1975. It is a pleasure to thank my hosts for their invitation, their hospitality, and for providing so discriminating an audience. I should also apologize for my delay in submitting this manuscript. In the intervening time some progress has been made with the theory, and I have taken the opportunity to mention some of it below. Moreover, a number of other sources have appeared, and of these [96] and [99] can be recommended as particularly useful to experienced topologists who want to see the results of the subject. However, my object has been a more elementary exposition, which I hope may convey the basic ideas of the subject in a way as nearly painless as I can make it. In this the Princeton audience encouraged me; the more I found means to omit the technical details, the more they seemed to like it. If that is the reaction of seasoned topologists, I hope that beginners may find it useful to have a gentle introduction to the ideas used in the current literature.

I am very grateful to J. P. May, B. J. Sanderson and S. B. Priddy for reading the first draft of this book, in part or in whole; I have benefited greatly from their comments. It goes without saying that I accept the responsibility for any jokes which remain.

J. F. ADAMS

TABLE OF CONTENTS

Infinite Loop Spaces

CHAPTER 1

BACKGROUND AND PRELIMINARIES

§1.1. *Introduction*

In this introductory chapter I will begin with some historical remarks. At the same time I will sketch in some background material, familiar enough to the specialists, but necessary if the nonspecialist is to have a fair chance of reading as far as he wishes. I will sketch the existence of the following three fields.

(i) The study of infinite-loop-spaces.

(ii) The study of stable homotopy theory, via spectra.

(iii) The study of generalized homology and cohomology theories.

I will also try to explain the very close relations which hold between these three topics. Finally I will give a rough classification or survey of the spaces presently known to be infinite-loop-spaces.

§1.2. *Loop-spaces*

In this section I will introduce loop-spaces.

Let X be a space, with base-point x_0. By the loop-space ΩX, I mean the function-space

$$(X,x_0,x_0)^{(I,0,1)}$$

of continuous functions $\omega : I \to X$ from the unit interval $I = [0,1]$ to the space X which carry 0 to x_0 and 1 to x_0. We give it the compact-open topology; as its base-point, if it needs one, we take the function ω_0 constant at x_0. The functions $\omega \in \Omega X$ are called loops in X.

Two historical references are compulsory at this point. First we have the well-known work of Marston Morse. In [111] Morse considered a Riemannian manifold M, say connected; and he sought information about the set of geodesics in M from the point P to the point Q. He found a relationship between this set of geodesics and the topology of the space of all paths from P to Q; by the latter we mean the function space

$$(M,P,Q)^{(I,0,1)}$$

of continuous functions $\omega : I \to M$ which carry 0 to P and 1 to Q. The homotopy type of this space is actually independent of the choice of P and Q, so this space is equivalent to ΩM. For example, suppose M is a sphere S^n, with its usual Riemannian structure; then one knows how many geodesics there are from P to Q. (More precisely, one may go from P to Q by the shortest geodesic, of length say θ; but by starting in the same direction and failing to stop when one might, one may reach Q by a geodesic of length $2n\pi + \theta$; and by starting in the opposite direction, one gets geodesics of length $2n\pi - \theta$.) From this Morse was able to calculate the homology groups $H_*(\Omega S^n)$ of the loop-space ΩS^n. Running his method in the other direction, he then deduced that if you took the sphere S^n, and gave it some other Riemannian structure, you would still have an infinity of geodesics from P to Q.

Secondly we have the well-known work of Serre. In [129], Serre generalized the theorem which says that there are infinitely many geodesics from P to Q, so as to replace the sphere S^n by any complete Riemannian manifold whose homology is not that of a point. However this result is only one of the good things in this paper, and perhaps some of the others were even more important; I have in mind particularly the methods which Serre introduced.

To help in understanding the loop-space ΩX, Serre introduced the path-space

$$EX = (X,x_0)^{(I,0)},$$

that is, the space of continuous functions $f : I \to X$ which carry 0 to x_0. The space EX is contractible, but it is a useful intermediate between ΩX and X. Serre defined a continuous function

$$p : EX \to X$$

by

$$p(f) = f(1) ;$$

thus p assigns to each path its end-point. He showed that this function p has the homotopy lifting property for maps of cubes; as we now say, it is a fibering in the sense of Serre. The fibre $p^{-1}x_0$ is exactly the loop-space ΩX. We say that we have a Serre fibering

$$\Omega X \longrightarrow EX \xrightarrow{p} X .$$

(Here I interpose a note on notation. Out of historical piety I am following Serre, who used the letter E. Some later authors write PX for the path-space.)

This sort of construction has great potential generality; for by using it, Bourbaki later showed that any continuous map $f : X \to Y$ can be replaced by a fibering in the sense of Serre. To do so you have to replace X by another space which is homotopy-equivalent to X, but you can keep the same space Y.

Serre also showed that the apparatus of spectral sequences, due to Leray, carried over to this context. The use of spectral sequences to do homological calculations gave Serre's methods great technical power, and they were very successful; one may add that Serre's exposition was lucid and elegant, and so it is not surprising that his methods were widely copied.

Before I go on, I must recall some other background material which has been known for a long time. Perhaps the first thing which a homotopy-theorist knows about loop-spaces is that they allow him to manipulate homotopy groups by moving them from one dimension to the next. More precisely, we have

$$\pi_i(\Omega X) \cong \pi_{i+1}(X) .$$

We can proceed more generally. Let W be a further space, with base-point w_0. Then maps

$$f : W \to X^I$$

are in (1-1) correspondence with maps

$$g : W \times I \longrightarrow X$$

in the following way:

$$(fw)(t) = g(w, t) \qquad (w \in W,\ t \in I) .$$

If we take account of the base-points, we find that maps

$$f : W, w_0 \longrightarrow \Omega X, \omega_0$$

are in (1-1) correspondence with maps

$$g : \Sigma W, \sigma_0 \longrightarrow X, x_0 .$$

Here ΣW is the quotient space obtained from $W \times I$ by identifying the subspace $(W \times 0) \cup (w_0 \times I) \cup (W \times 1)$ to a single point, which becomes the base-point σ_0 in ΣW. This quotient space is called the reduced suspension of W; it is often written SW.

Anyway, passing to homotopy classes we find a natural (1-1) correspondence

(1.2.1) $$[W, \Omega X] \longleftrightarrow [\Sigma W, X] .$$

Here I write $[U,V]$ for the set of homotopy classes of maps from U to V, where both maps and homotopies are supposed to preserve the base-point.

We would now express this by saying that the functors Σ and Ω are "adjoint." Of course this terminology is more recent, being due to Kan [75].

In particular we can take W to be the quotient space $I^n/\partial I^n$, where I^n is the unit cube in R^n and ∂I^n is its boundary. We get

$$\Sigma W = \frac{I^n \times I}{(I^n \times 0) \cup (\partial I^n \times I) \cup (I^n \times 1)} = \frac{I^{n+1}}{\partial I^{n+1}},$$

and our (1-1) correspondence becomes

$$\pi_n(\Omega X) \cong \pi_{n+1}(X). \tag{1.2.2}$$

Of course, I have yet to explain why homotopy-theorists should want to move homotopy groups from one dimension to the next; this belongs to the next section.

§1.3. *Stable homotopy theory*

In this section I will introduce stable homotopy theory and spectra.

Topologists make a basic distinction between stable phenomena and unstable phenomena; a phenomenon is said to be stable if it can occur in any dimension, or any sufficiently large dimension, in a way which is essentially independent of the dimension. The construction which varies the dimension is usually suspension. For example, in homology-theory we have

$$\tilde{H}_n(W;\pi) \cong \tilde{H}_{n+1}(\Sigma W;\pi),$$

where $\tilde{H}$ means reduced homology. But the principle is to be seen more clearly in homotopy-theory, where it goes back to Freudenthal [59]. For our purposes, let W and X be complexes of some sufficiently good sort, say CW-complexes. Then the suspension construction gives a function

$$\Sigma : [W,X] \longrightarrow [\Sigma W, \Sigma X];$$

and we have the following well-known theorem.

THEOREM 1.3.1. *Suppose that* X *is* (n–1)*-connected and* W *is of dimension* d; *then the function*

$$\Sigma : [W,X] \longrightarrow [\Sigma W, \Sigma X]$$

is onto if $d \leq 2n-1$, *and is a* (1–1) *correspondence if* $d \leq 2n-2$.

A suitable textbook reference is [135], especially p. 458.

To prove this theorem, we replace $[\Sigma W, \Sigma X]$ on the right by $[W; \Omega\Sigma X]$, using (1.2.1). We obtain the following commutative diagram.

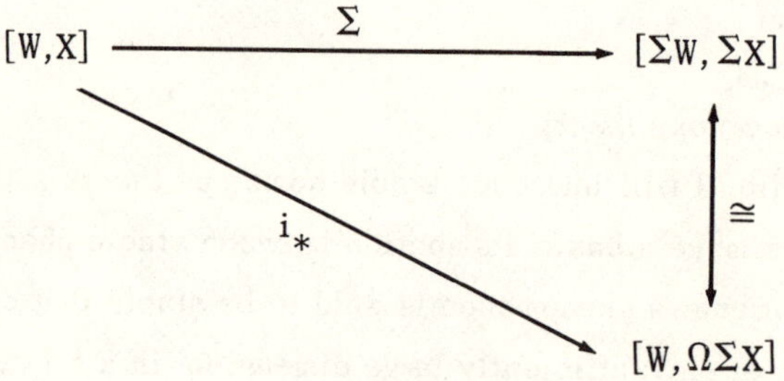

Here i_* means the function induced by the map i, and $i : X \to \Omega\Sigma X$ is the map corresponding under (1.2.1) to the identity map $1 : \Sigma X \to \Sigma X$. This diagram allows one to prove the theorem by studying the loop-space $\Omega\Sigma X$ and its relation with X; similarly for other theorems in suspension-theory.

In fact, the study of loop-spaces proved a most valuable method in homotopy-theory. I. M. James [73] succeeded in replacing the rather large function-space ΩS^n by an explicit cell-complex which was so small that its structure could be well understood; I shall call this the "James model" for ΩS^n. It led him to good new results in suspension-theory. Bott and Samelson [44] succeeded in computing the homology of a number of loop-spaces, including the loop-space $\Omega(S^{n_1} \vee S^{n_2} \vee \cdots \vee S^{n_d})$ on a wedge-sum or "bouquet" of spheres. Building on this, P. J. Hilton [63] succeeded in analyzing the homotopy type of $\Omega(S^{n_1} \vee S^{n_2} \vee \cdots \vee S^{n_d})$; this again led to results in homotopy-theory.

Let us return to generalities. With the notation of (1.3.1), the homotopy classification problem of determining $[W, X]$ is said to be "stable" if $d \leq 2n-2$, so that we meet exactly the same problem for $[\Sigma W, \Sigma X]$, for $[\Sigma^2 W, \Sigma^2 X]$ and so on in higher dimensions. More generally, we may say that stable homotopy-theory is the part of homotopy-theory which studies phenomena which are stable in the intuitive sense described above. In order to persuade unbelievers, one has to show that it contains theorems of interest.

A topic which was studied quite early, and which provided a good advertisement for stable homotopy-theory, was Spanier-Whitehead duality [137, 134]. Suppose we have a good space X, say a finite complex, and we embed X in a sphere S^n in a way which is not pathological, so that the complement $\mathcal{C}X$ of X in S^n has a finite complex Y as a deformation retract, and similarly $\mathcal{C}Y$ has X as a deformation retract. Then the Alexander duality theorem tells us that the homology and cohomology of Y are determined by X and do not depend on the embedding of X in S^n. On the other hand the fundamental group $\pi_1(Y)$ is not determined by X and does depend on the embedding; it is sufficient to consider the example $X = S^1$, $n = 3$ (classical knots).

The question then arises, how much about Y is determined by X? The answer is that X determines the stable homotopy type of Y. Here we say that Y and Z are of the same stable homotopy type if there exists an integer m such that $\Sigma^m Y$ and $\Sigma^m Z$ are homotopy-equivalent. Alternatively, we can define a new category, the stable homotopy category of Spanier and Whitehead [136, 138], by saying that the objects shall be the finite complexes, and $\{Y, Z\}$, the set of stable homotopy classes of maps from Y to Z, shall be

$$\lim_{m \to \infty} [\Sigma^m Y, \Sigma^m Z] .$$

This limit is attained by (1.3.1); if Y were not finite-dimensional this definition of $\{Y,Z\}$ would not be appropriate. Then Y and Z are of the same stable homotopy type exactly when they are equivalent in the stable homotopy category.

The question about embeddings in S^n may now be answered more explicitly by saying that Y depends on X via a contravariant functor $D = D_n$. Here the functor $D = D_n$ takes values in the stable homotopy category of Spanier and Whitehead; it is defined on a full subcategory of that, for it is defined on all objects X which can be well embedded in S^n, and on all morphisms in the category between such objects X. Spanier and Whitehead call D_nX the "n-dual" of X.

However, it was soon observed that the stable homotopy category constructed by Spanier and Whitehead does not contain sufficient objects for some purposes (even if we relax the assumption that our complexes are finite). A compelling example was provided by Thom's work on cobordism [151]. He reduced the study of cobordism groups to the study of certain stable homotopy groups

$$\lim_{n \to \infty} \pi_{n+r}(MO(n))$$

$$\lim_{n \to \infty} \pi_{n+r}(MSO(n))$$

and so on. Here the spaces $MO(n)$, $MSO(n)$ etc., are ones constructed by Thom, and usually called "Thom complexes"; the limits $\lim_{n \to \infty}$ can be formed because the spaces $MO(n)$, $MSO(n)$ and so on come provided with maps

$$\Sigma MO(n) \longrightarrow MO(n+1)$$

$$\Sigma MSO(n) \longrightarrow MSO(n+1)$$

and so on.

It was explicitly stated by John Milnor ([104], especially pp. 511-512) that one could argue very much more clearly if one could work in a category where there is a single object **MO** rather than the spaces MO(n) which approximate to it, and similarly a single object **MSO**, and so on. Now such a context was already known [81, 82]. For our purposes, a "spectrum" E is a sequence of spaces E_n (with base-point) provided with maps $\varepsilon_n : \Sigma E_n \to E_{n+1}$. For example, the spaces MO(n) provided with their maps $\Sigma MO(n) \to MO(n+1)$ constitute a spectrum, the Thom spectrum **MO**. Similarly for **MSO**. As another example, let X be a CW-complex with base-point; its "suspension spectrum" is the spectrum in which the nth space is $\Sigma^n X$ and the maps are the identity maps $\Sigma(\Sigma^n X) = \Sigma^{n+1} X$. We shall write this spectrum $\Sigma^\infty X$; we would like to say that this defines a functor Σ^∞ from CW-complexes to spectra. Of course, for this purpose (and for other purposes) we have to make spectra into a category. One has to get the notion of a "map of spectra" right, in a way that is slightly technical. However, it is now generally agreed and understood that the best category in which to do stable homotopy theory is the category of spectra constructed by J. M. Boardman [34, 35, 36, 153], or else some category equivalent to Boardman's. I have tried to give an exposition in the most elementary terms in [9], especially pp. 131-146. For the moment I summarize the following points from it.

(i) A CW-spectrum is one in which each space E_n is a CW-complex (with base-point) and each map $\varepsilon : \Sigma E_n \to E_{n+1}$ embeds ΣE_n as a subcomplex of E_{n+1}. One can take the CW-spectra as the objects of the required category.

(ii) Σ^∞ is a functor; and whenever X is finite-dimensional it induces a (1-1)-correspondence

$$\lim_{n\to\infty}[\Sigma^n X, \Sigma^n Y] \longrightarrow [\Sigma^\infty X, \Sigma^\infty Y] .$$

(Here $[\Sigma^\infty X, \Sigma^\infty Y]$ means homotopy classes of maps in the category of CW-spectra.)

(iii) The category of CW-spectra is so arranged that one can carry over all the constructions which one usually performs with CW-complexes.

(iv) If you like you can make the category of CW-spectra into a graded category. Let E be a spectrum; then by reindexing it we can get a new spectrum; for example, we can define F by $F_n = E_{n+1}$. Then a map of degree 1 from E to G will be an ordinary map (of degree 0) from F to G. We write $[X,Y]_n$ for the set of homotopy classes of maps of degree n from X to Y.

This is sufficient to give the basic ideas. The important thing is to know that there is a good category of spectra, and not to insist on any one choice of details for its construction. In fact there are alternative ways of setting up the details; they all lead to the good category, but for some particular application one may have an advantage over another. Let us keep our options open.

This completes the introduction of stable homotopy theory and spectra; we are now ready to go back and talk about loop-spaces again.

§1.4. *Infinite loop spaces*

In this section I will introduce infinite loop spaces.

A loop-space is better than an ordinary space; not every space is equivalent to a loop-space. For example, a loop-space is an H-space, and not every space is an H-space. Here I recall that a space X is an H-space if it comes equipped with a product map

$$\mu : X \times X \longrightarrow X$$

which satisfies appropriate axioms. Equivalently, we may suppose given for each W a product operation on the set $[W,X]$, so that this product is natural for maps of W. The minimum axiom is that the map from W to X which is constant at the base-point should act as a unit for this product; equivalently, the base-point in X should act as a unit (up to homotopy) for the map μ. (All maps and homotopies are supposed to preserve the base-point.)

It is clear that a loop-space $X = \Omega Y$ is an H-space. In fact, we can give the product map $\mu : \Omega Y \times \Omega Y \to \Omega Y$ explicitly and directly by

$$(\mu(\omega', \omega''))(t) = \begin{cases} \omega'(2t) & (0 \leq t \leq \tfrac{1}{2}) \\ \omega''(2t-1) & (\tfrac{1}{2} \leq t \leq 1) . \end{cases}$$

That is, the product loop spends the first half of its time going along ω' at double speed, and the second half of its time going along ω''. Alternatively, the set $[W, \Omega Y]$ becomes $[\Sigma W, Y]$, or equivalently the fundamental group of the function space

$$(Y, y_0)^{(W, w_0)} ,$$

so that $[W, \Omega Y]$ is a functor from spaces W to groups.

The concept of "H-space" goes back to Serre [129], who chose the letter H in honor of Hopf's work on the topology of Lie groups. H-spaces are now the subject of an extensive literature. The remark that a loop-space is an H-space is also due to Serre.

If X is an H-space, we can use the structure map $\mu : X \times X \to X$ to define a "Pontryagin product" of homology classes in $H_*(X)$ [44]. This piece of structure is vital to understanding the homology of H-spaces such as ΩS^n and $\Omega(S^{n_1} \vee S^{n_2} \vee \cdots \vee S^{n_d})$. More precisely, $H_*(\Omega S^n)$ is a free algebra (over Z) on one generator of degree $n-1$, and similarly $H_*(\Omega(S^{n_1} \vee S^{n_2} \vee \cdots \vee S^{n_d})$ is a free algebra (over Z) on generators of degree $n_1-1, n_2-1, \cdots, n_d-1$. Here "algebras" are supposed to be associative, but not necessarily commutative.

We can of course iterate the loop-space construction, forming

$$\Omega^2 X = \Omega(\Omega X) = (X, x_0)^{(I^2, \partial I^2)} ,$$

and so on. If a loop-space is an unusually good space, then a double loop-space is even better; not all loop-spaces are double loop-spaces.

I will say that X is an "infinite loop space" if there is a sequence of spaces $X_0, X_1, X_2, \cdots$ with $X_0 = X$ and with weak equivalences

$$X_n \xrightarrow{\simeq} \Omega X_{n+1} .$$

Here I recall that a map $f: X \to Y$ between connected spaces is a weak equivalence if

$$f_*: \pi_r(X) \longrightarrow \pi_r(Y)$$

is an isomorphism for all r; this implies that

$$f_*: [W,X] \longrightarrow [W,Y]$$

is an isomorphism for all CW-complexes W. If X and Y are not connected we first demand that

$$f_*: \pi_0(X) \longrightarrow \pi_0(Y)$$

is a (1-1) correspondence, and then legislate as above for each path-component.

Weak equivalences are used to avoid technical nuisance. For example, if one merely asserts that ΩS^n and the "James model" for it are weakly equivalent, then one avoids having to prove that they are actually homotopy-equivalent, but one still gets all the theorems one is interested in.

Let us return to our sequence of spaces $X_0, X_1, X_2, \cdots$. Each weak equivalence

$$X_n \longrightarrow \Omega X_{n+1}$$

can of course be transformed by (1.2.1) into a map

$$\Sigma X_n \longrightarrow X_{n+1} .$$

So such a sequence of spaces X_n is a spectrum of some sort. We will introduce a name for this sort of spectrum. Let E be a spectrum, and let us take the given structure maps

$$\varepsilon_n : \Sigma E_n \longrightarrow E_{n+1}$$

and transform them into maps

$$\varepsilon'_n : E_n \longrightarrow \Omega E_{n+1} ;$$

we say that E is an "Ω-spectrum" if the maps ε'_n are weak equivalences. If E is an Ω-spectrum we have the following special property:

$$[\Sigma^\infty X, E]_n = [X, E_n] .$$

In any case, we may say that an infinite loop space X is the 0th term of an Ω-spectrum $\mathbf{X} = \{X_n\}$.

If it helps, we can replace the spaces X_n by weakly equivalent ones so that we actually get homeomorphisms $X_n \cong \Omega X_{n+1}$ [89]. In fact, there are some arguments for defining an "Ω-spectrum" so that the maps

$$\varepsilon'_n : E_n \longrightarrow \Omega E_{n+1}$$

are required to be homeomorphisms rather than weak equivalences; this is very appropriate in the theory of infinite loop spaces, where it is necessary in order to keep the geometry under control and keep the details accurate. We can still construct the good category of spectra by taking as objects the "Ω-spectra" in this sense. But as my immediate purpose is to explain the relation with homotopy-theory, the point is not important just yet.

I should now give some examples of infinite loop spaces. The first example is provided by the Eilenberg-MacLane spaces. Let π be an abelian group, and let X_n be an Eilenberg-MacLane complex of type (π, n), so that its homotopy groups are given by

$$\pi_r(X_n) = \begin{cases} \pi & \text{if } r = n \\ 0 & \text{if } r \neq n . \end{cases}$$

Using (1.2.2), we see that ΩX_{n+1} is also of type (π, n); so there is a weak equivalence $X_n \to \Omega X_{n+1}$. Thus any Eilenberg-MacLane space is an infinite loop space.

The second example is provided by $Z \times BU$; here BU may be viewed either as the classifying space of the "infinite-dimensional unitary group" $U = \bigcup_n U(n)$, or as the limit of classifying spaces $\lim_{n \to \infty} BU(n)$. According to the Bott periodicity theorem [42, 43, 50, 18] we have a weak equivalence

$$Z \times BU \simeq \Omega^2(Z \times BU) ;$$

so $Z \times BU$ is an infinite loop space. Similarly for $Z \times BO$.

Both these examples may be obtained from generalized cohomology theories, but that topic belongs to the next section.

§1.5. *Generalized cohomology theories*

In this section I will introduce generalized cohomology theories.

I assume it is more or less known that a generalized homology or cohomology theory is a functor which satisfies the first six axioms of Eilenberg-Steenrod, but not the seventh or "dimension" axiom. When we need it we may throw in the wedge axiom of Milnor [106]. My homology and cohomology functors will be defined in the first instance on CW-complexes.

We may also grant that the study and application of such functors is of interest to topologists. The ones most often useful are of three sorts.

(i) Ordinary or classical homology and cohomology.

(ii) The many variants of K-theory.

(iii) The many variants of bordism and cobordism.

It is probably as well to give particular attention to the case in which the covariant functor is stable homotopy. In a sense we have listed it already, for it occurs under (iii) above in the guise of framed bordism; but homotopy-theorists regard it as more elementary than that, and especially important to their subject. We may define the stable homotopy groups of a complex X by

$$\pi_n^S(X) = \{S^n, X\} = \lim_{m\to\infty} [S^{m+n}, \Sigma^m X]$$

$$= [\Sigma^\infty S^0, \Sigma^\infty X]_n \, .$$

We may define the homotopy groups of a spectrum (which are automatically "stable") by

$$\pi_n(X) = [\Sigma^\infty S^0, X]_n \, .$$

It turns out that these functors do satisfy the axioms for a homology theory. The corresponding contravariant functor is stable cohomotopy. If we think say of stable homotopy, this functor occupies an extreme position; one may make this formal by showing that this functor is initial among theories of a suitable sort; informally, we may agree that this functor contains a great deal of information and is maximally hard to compute.

§1.6. *The relation between spectra and generalized cohomology theories*

I will now begin to relate some of the notions I have introduced.

Let us start by supposing given a generalized cohomology theory k^*. It can certainly be made to do the following things.

(i) To each space X with base-point x_0 and each integer n it assigns a (reduced) cohomology group $\tilde{k}^n(X)$. (For this purpose the word "space" means a CW-complex.)

(ii) To each map $f: X,x_0 \to Y,y_0$ it assigns an induced homomorphism

$$f^*: \widetilde{k}^n(X) \longleftarrow \widetilde{k}^n(Y) .$$

(iii) To each space X with base-point x_0 it assigns an isomorphism

$$\sigma: \widetilde{k}^n(X) \xrightarrow{\cong} \widetilde{k}^{n+1}(\Sigma X) .$$

These things $\widetilde{k}^n(\ )$, f^* and σ have properties which may be inferred from the Eilenberg-Steenrod axioms (if one prefers that starting-point) plus Milnor's wedge axiom. Alternatively, we can write an axiomatic system of cohomology theory in terms of these things $\widetilde{k}^n(\ )$, f^* and σ, with their usual properties as axioms. The precise properties in question need not detain us here, but they permit one to apply the Representability Theorem of E. H. Brown [48, 49]; this theorem asserts that a contravariant functor from CW-complexes to sets which satisfies certain hypotheses is of the form $[\ \ ,Y]$.

If you will allow me a slight oversimplification which I will correct later, we see that there exist CW-complexes Y_n and (1-1) correspondences

$$\widetilde{k}^n(X) \longleftrightarrow [X,Y_n]$$

defined when X is a CW-complex and natural for maps of X. But now we have the following composite (1-1)-correspondence.

$$\begin{array}{ccc}
[X,Y_n] & & [X,\Omega Y_{n+1}] \\
 & & \downarrow \cong \\
\updownarrow \cong & & [\Sigma X, Y_{n+1}] \\
 & & \downarrow \cong \\
\widetilde{k}^n(X) & \xrightarrow[\cong]{\sigma} & \widetilde{k}^{n+1}(\Sigma X)
\end{array}$$

This can only happen if we have a weak equivalence

$$Y_n \xrightarrow{\simeq} \Omega Y_{n+1} .$$

So any generalized cohomology theory yields an Ω-spectrum. We may arrange matters so that this Ω-spectrum lies in a good category of spectra.

The oversimplification consists in omitting to mention that when we apply the theorem of E. H. Brown, we must run X over connected complexes only. It is easy to deal with this point at the price of complicating the argument a little; for a more complete account, see [9], pp. 131-134.

As an example of the general construction considered above, suppose we take k^* to be ordinary cohomology with coefficients in π,

$$k^n(X) = H^n(X;\pi) ;$$

then we have

$$\widetilde{k}^n(X) = \widetilde{H}^n(X;\pi) = [X,Y_n] ,$$

where Y_n is an Eilenberg-MacLane complex of type (π,n). The Ω-spectrum which we obtain is the Eilenberg-MacLane spectrum for the group π.

Similarly, taking k^* to be complex K-theory, we obtain a Ω-spectrum in which every even term is the space $Z \times BU$ and every odd term is the space U. Similarly for real K-theory and $Z \times BO$.

We can also proceed in the reverse direction. In a celebrated paper [155], G. W. Whitehead showed that if we start from any spectrum E, we can define a generalized homology theory and a generalized cohomology theory. This is done in such a way that the Thom spectrum **MO**, although it is not an Ω-spectrum, gives rise to real, unoriented bordism and cobordism; the Thom spectrum **MSO** gives rise to real, oriented bordism and cobordism; the sphere spectrum $\Sigma^\infty S^0$ gives rise to stable homotopy and cohomotopy; and so on.

The good definition of E-cohomology, valid for any CW-complex X, is

$$\widetilde{E}^n(X) = [\Sigma^\infty X, E]_{-n} \,.$$

Of course this extends instantly to a definition of E-cohomology of spectra; if X is a spectrum we define

$$\widetilde{E}^n(X) = [X, E]_{-n} \,.$$

The definition of E-homology which you would first think of would be modelled on Alexander duality; if X is a finite complex, try

$$\widetilde{E}_q(X) = \widetilde{E}^{n-q-1}(Y) \,,$$

where Y is a Spanier-Whitehead dual of X in S^n. However it is profitable to manipulate this definition into a form which can be generalized. One form of the definition, valid for any CW-complex X, is

$$\widetilde{E}_q(X) = \lim_{n\to\infty} \pi_{n+q}(E_n \wedge X) \,.$$

Here the "smash product" $W \wedge X$ is defined by

$$W \wedge X = \frac{W \times X}{(W \times x_0) \cup (w_0 \times X)} \,,$$

where w_0 and x_0 are the base-points in W and X.

The coefficient groups of a cohomology theory k^* are given by k^n of a point, that is $\widetilde{k}^n(S^0)$; so for the cohomology theory obtained from E the coefficient groups are

$$\widetilde{k}^n(S^0) = \widetilde{E}^n(S^0) = [\Sigma^\infty S^0, E]_{-n} = \pi_{-n}(E) \,,$$

the homotopy groups of E. (Similarly for the homology theory.) This is also the equation one gets if one starts from a cohomology theory k^* and constructs a representing spectrum E.

Perhaps this is the right point to issue one warning. The coefficient groups $k^n(S^0)$ of a generalized cohomology theory can be non-zero for many values of n, both positive and negative; for example, taking the cohomology theory in question to be complex K-theory, we get the group Z when n is even, whether n is positive, negative or zero. The same remark therefore applies to the homotopy groups of the representing spectrum. In particular, if we were hoping to apply the Hurewicz theorem, we may be stuck; a spectrum X need not have any dimension d for which we can assert $\pi_i(X) = 0$ for $i < d$. A spectrum X is said to be "bounded below" if there exists d such that $\pi_i(X) = 0$ for $i < d$. Some authors say that X is "connective" if $d = 0$ will do, so that $\pi_i(X) = 0$ for $i < 0$.

To continue, we have said that from any cohomology theory we can construct a spectrum, and from any spectrum we can construct a cohomology theory. In order to convince ourselves that the topics of spectra and of cohomology theories are essentially equivalent, we must see that these constructions are essentially inverse.

If we start from a cohomology theory, construct a representing spectrum, and take the corresponding cohomology theory, then we recover the original cohomology theory up to isomorphism.

Conversely, suppose we start from a spectrum E, construct the corresponding generalized cohomology theory E^*, and construct the representing Ω-spectrum F. Then E and F have the same homotopy groups (namely the coefficient groups for E^*); and it can be shown that one can choose an equivalence between E and F.

(It may be objected that the above account is inadequate, because it concentrates on objects and neglects morphisms; I "ought" to be showing that the two constructions are functorial. However this would involve questions which I want to omit so that we can get on to something else.)

§1.7. *The relation between spectra and infinite loop spaces*

I will continue to relate some of the notations I have introduced.

We have just seen that any spectrum E can be replaced by an equivalent Ω-spectrum F. In fact the direct construction of F from E would be perfectly possible; we could take

$$F_n = \lim_{m \to \infty} \Omega^m E_{n+m} .$$

But let us use the construction at the end of §1.6, and define $\Omega^\infty E$ to be F_0, the 0th term of the Ω-spectrum representing E^*. Then we have the following composite (1-1) correspondence:

$$\begin{array}{ccc} [X, \Omega^\infty E] & = & [X, F_0] \\ & & \Big\updownarrow \cong \\ [\Sigma^\infty X, E] & = & \widetilde{E}^0(X) \end{array}$$

Thus Ω^∞ is a functor from spectra to spaces, adjoint to the functor Σ^∞ from spaces to spectra. The values of the functor Ω^∞ are the infinite loop spaces.

It will be objected that some authors use the symbol Ω^∞ for something else. Such authors will concede that I have a genuine need for the two adjoint functors, one from spaces to spectra and one from spectra to spaces, which I propose to call Σ^∞ and Ω^∞. In turn I will concede that what I am calling Ω^∞ can be factored into two functors, (i) the functor which converts any spectrum E into an equivalent Ω-spectrum F, and (ii) the functor which passes from the Ω-spectrum F to the space F_0. For our present purpose the more important of these two steps is (ii), the passage from spectra to spaces. As for the functor of step (i), even though some call it Ω^∞, I would prefer to see it kept in its proper place, which is inside a black box.

It may seem that the remarks above are not enough to establish any close connection between spectra and infinite loop spaces; after all, the functor Ω^∞ may lose information. However, when an author says "the space X is an infinite loop space," he usually means that he has constructed a specific Ω-spectrum $\mathbf{X}$ whose 0th term is X. In this case the reader may well feel the need of more information about that Ω-spectrum; which of the many inequivalent Ω-spectra with the same 0th term has the author got his grubby hands on – for you may be sure there is more sense in his proof than in his statement? Equivalently, the author says "this functor $\tilde{k}^0$ is the 0th term of a generalized cohomology theory"; he must have constructed such a cohomology theory – how can we tell which it is? Again, the author says "the spaces X and Y are infinite loop spaces, but I do not know if the map $f: X \to Y$ is an infinite loop map." He means that he has constructed spectra $\mathbf{X}$ and $\mathbf{Y}$ such that $X = \Omega^\infty \mathbf{X}$ and $Y = \Omega^\infty \mathbf{Y}$, but he does not know if there is a map of spectra $\mathbf{f}: \mathbf{X} \to \mathbf{Y}$ *between the spectra he has constructed* such that $f = \Omega^\infty \mathbf{f}$. How are we to help him unless we can describe the spectra $\mathbf{X}$ and $\mathbf{Y}$ in terms that we can work with?

One thing that can be done to palliate the situation in practice is to make the functor Ω^∞ retain more information. In particular, let $\mathbf{X}$ be an Ω-spectrum; then the space ΩX_1 (being a loop-space) has an H-space structure; using the equivalence $X_0 \xrightarrow{\simeq} \Omega X_1$, we get an H-space structure on X_0. (For example, in the case of the complex K-cohomology theory, the H-space structure on $Z \times BU$ is that corresponding to the Whitney sum of vector-bundles; similarly for real K-theory and $Z \times BO$.) Anyway, we can arrange that the functor Ω^∞ takes values in the category of H-spaces, instead of the category of spaces. This means that the functor Ω^∞ loses less information. It is a step in the right direction, but it leaves us a long way to go. The procedure usually adopted is to make the functor Ω^∞ take values in a category of spaces with so much extra structure that the functor Ω^∞ loses no information at all. (I will sketch this procedure in Chapter 2.) In this sense, the study of infinite loop spaces is substantially equivalent to the study of spectra.

From this point of view, it might appear a matter of taste and convenience whether one works with properties or invariants usually thought of as belonging to the spectrum X or to the space $\Omega^\infty X$. Here of course I mean invariants of $\Omega^\infty X$ which reflect the infinite-loop-structure, such as the homology operations of Kudo and Araki [77, 78], Dyer and Lashof [55], and such as the transfer to be considered in Chapter 4. Just now I want to stress that it is useful to have both approaches available.

For example, suppose we have a map of connective spectra $f\colon X \to Y$ and we wish to prove that the induced map of homotopy groups $f_*\colon \pi_*(X) \to \pi_*(Y)$ is epi. It will be sufficient to prove the corresponding result for

$$(\Omega^\infty f)_*\colon \pi_*(\Omega^\infty X) \longrightarrow \pi_*(\Omega^\infty Y)\ ;$$

and now it is perfectly possible that we may find a map $g\colon \Omega^\infty X \longleftarrow \Omega^\infty Y$ of spaces, which is not an infinite loop map, such that

$$(\Omega^\infty f)g \simeq 1\colon \Omega^\infty Y \longrightarrow \Omega^\infty Y\ .$$

This will prove the result; and this suggestion is not artificial, for this is exactly what happens in the case of the Kahn-Priddy theorem [74]. (See §4.1 for more detail about the Kahn-Priddy theorem, and see [8] for a commentary on it bearing on the present point.) Therefore, we may still make a profit by playing off unstable homotopy theory against stable homotopy theory, and by looking at the same problem from more than one point of view. To say it another way: if we want information, we must be prepared to go wherever the geometry is.

§1.8. *Survey of examples*

I will now give a rough classification of the spaces currently known to be infinite loop spaces, under three main headings. Of course these headings overlap, and their interrelations will provide material for later chapters. But broadly, the spaces currently known to be infinite loop spaces are of three sorts.

(i) Those which arise in the way described in §1.6 from known generalized cohomology theories, such as ordinary cohomology, K-theory, cobordism and their variants.

(ii) Those which can be constructed by stable homotopy theory, using various constructions on spectra and then applying Ω^∞.

(iii) Those which can be constructed by the machinery to be considered in Chapter 2.

The machinery to be considered in Chapter 2 is usually adjusted so as to turn out connective spectra E; so if the descriptions below fail to specify the coefficient groups $\pi_n(E)$ for $n < 0$, it doesn't matter, because these groups are zero.

To continue, three subdivisions of (iii) account for the examples of most interest.

(iii a) Those related to the geometry of manifolds.

(iii b) Those related to groups of units in cohomology rings.

(iii c) Those related to algebraic K-theory.

I begin with (iii a). When one considers piecewise-linear manifolds and topological manifolds, one has to introduce appropriate versions of bundle-theory. The stable bundle-theories are representable functors with representing spaces BPL, BTop. The existence of BPL is usually attributed to Milnor [105] but a more accessible reference is [110]. The existence of BTop is implicit in [108]. One can also introduce a space, variously called BF, BG or even BH, which corresponds to (the stable version of) a very homotopy-theoretic sort of bundle-theory, in which one studies fibre-homotopy-equivalence classes of fiberings with fibres homotopy-equivalent to spheres [140]. The corresponding "groups" PL, Top and F exist in suitable senses, and so do "coset spaces" such as F/PL and PL/O, but it is not necessary to insist on the details here.

The immediate point is that the spaces BPL, BTop and BF are infinite loop spaces. This result was announced by Boardman and Vogt in [39]; a full-dress account of their work appeared in [40], but there the authors chose not to cover the case of BPL (see [40] pp. 216-217). For

BPL the reader may refer to May [99]. Moreover, the statement that these spaces are infinite loop spaces needs to be amplified; for example, the H-space structure on each of BPL, BTop and BF corresponds to the Whitney sum of "bundles." We may say then that the PL K-group $K_{PL}(X)$ is the 0th term of a cohomology theory; similarly for $K_{Top}(X)$ and $K_F(X)$. It is also good to amplify the theorem by including results on the coset spaces such as F/PL; and finally it is good to state that certain maps, such as the canonical map $BPL \to BTop$, are infinite loop maps – that is, they lie in the image of the functor Ω^∞ (see §1.7).

Similar remarks apply to the "special" analogues SPL, STop and SF of PL, Top and F. However, one can go further. Suppose given a generalized cohomology theory k^*, and (say) a vector bundle ξ over X with total space E; and let E_0 be the complement of the zero-section. By an *orientation* of ξ over k^* we mean an element

$$u \in k^*(E, E_0)$$

such that when we restrict it to any fibre F of E, the restriction

$$i^*u \in k^*(F, F_0)$$

is a generator. Of course we have to say what we mean by a "generator," and we have to put enough assumptions on k^* to do so. But assuming all that, there is a bundle-theory in which we consider vector bundles ξ equipped with given orientations u over k^*. Similarly if we consider fiberings more general than vector-bundles. In terms of such oriented fiberings we construct a K-group; we need to state and prove that this is the 0th term of a cohomology theory, just as for K_{PL} and K_{Top}. The reader will find it done in [99].

Since an orientation is a necessary point of departure for many cohomological constructions, we must expect these K-groups to be the natural domain of definition of many invariants.

Next I seek to work in the direction of (iii b). Many of the transformations which we use in studying the geometry of manifolds transform addition into multiplication. For example, consider the total Chern class $c(\xi)$. It is defined on ξ if ξ is a U(n)-bundle over a space X, or even if ξ is an element of the group K(X); and it satisfies

$$c(\xi \oplus \eta) = (c\xi) \cdot (c\eta) .$$

If we want to construct a group to receive the values of c, it is natural to take the set of formal series

$$1 + x_2 + x_4 + x_6 + \cdots ,$$

where $x_{2q} \in H^{2q}(X)$, and make this set into a group G(X) by using the product of formal series. Then the total Chern class gives a homomorphism of groups

$$c\colon K(X) \longrightarrow G(X) .$$

It has been shown by Segal [128] that this group G(X) is the 0th term of a generalized cohomology theory. Actually Segal's result is more general; he considers formal series

$$1 + x_1 + x_2 + x_3 + \cdots$$

with

$$x_i \in H^i(X; A_i) ;$$

but it is restricted to ordinary cohomology.

However, one can ask a similar question in greater generality. Suppose given a generalized cohomology theory k^* which has cup-products; for example, K-theory and cobordism have such products. Then we can form the multiplicative group G(X) of elements

$$1 + x, \qquad x \in \tilde{k}^0(X) .$$

And we can ask if this is the 0th group of a cohomology theory.

For example, stable cohomotopy is a generalized cohomology theory with products (corresponding to the sphere spectrum $\Sigma^\infty S^0$); the corresponding multiplicative theory should be K_F.

In general, there is no good reason to suppose that $G(X)$ is the 0th group of a cohomology theory. More particularly, R. Steiner [143] has shown that complex K-theory with mod p coefficients provides a counterexample. It is therefore necessary to assume that the cup-product is especially good. For example, take k to be real K-theory, KO; then as a set $G(X)$ is represented by the usual space BO, considered as the subspace

$$1 \times BO \subset Z \times BO .$$

However, to represent $G(X)$ as a group we have to put an unusual H-space structure on BO; that is, we take a product map

$$\mu: BO \times BO \longrightarrow BO$$

defined using the tensor product of vector bundles; strictly speaking, it represents the tensor product of virtual bundles of virtual dimension 1. (A virtual bundle is the formal difference $\xi - \eta$ of two honest bundles ξ, η). We call this H-space $BO_\otimes$; similarly for $BU_\otimes$. It is known that these H-spaces $BO_\otimes$, $BU_\otimes$ are infinite loop-spaces [127]. A general result, valid for all cohomology theories with sufficiently good products, is given in [99].

All considerations of (iii c), the spectra of algebraic K-theory, is postponed to §2.6, §3.2.

This essentially completes my introduction. The theory of infinite loop spaces should have as its object to provide topologists with the information they need about infinite loop spaces – or equivalently spectra – or equivalently generalized cohomology theories, with special emphasis on those infinite loop spaces which arise in nature and are needed for the applications, particularly the applications to the theory of manifolds. It will take us five chapters to survey some of the tools of this trade. I shall return to a survey of the objects of the exercise and of its relative success in a short final chapter, Chapter 7.

CHAPTER 2

MACHINERY

§2.1. *Introduction*

The object of this chapter is to survey in somewhat greater detail the project mentioned in §1.7: to set up a category of spaces with so much extra structure that the functor Ω^∞ yields an equivalence from the category of spectra to this new category of structure-laden spaces. The apparatus of definitions, theorems and proofs needed to carry out this programme in detail demands a capital investment of intellectual work which may seem daunting to those not directly concerned; many readers may be able to remember feeling the same way about spectral sequences, sheaf-theory or whatever is now their favorite tool; let us be glad we don't work in algebraic geometry. Topologists commonly refer to this apparatus as "machinery."

In §2.2 and §2.3 I will try to show the necessity of the approach which is used to deal with these structure-laden spaces. This amounts to motivation for the definitions. In the course of §2.3 I shall transfer attention from definitions to theorems, and in §2.4 we will turn to methods of proof; but we will continue to survey machinery for §2.5 and §2.6. In §2.7 I will supplement the remarks about "additive" structures in §2.2-§2.6 with the remark that one should also consider "multiplicative" structures.

§2.2. *Loop-spaces and* A_∞ *spaces in the sense of Stasheff*

In this section we study the question: how does a homotopy-theorist tell whether a space X is equivalent to a loop-space ΩY?

First, the candidate X must come as an H-space. However, a loop-space is better than the general run of H-spaces. To begin with, a loop-space is equivalent to a topological monoid or semigroup – that is, to an H-space in which the product is strictly associative, and the unit is a strict unit. To prove this, we use the device of "Moore loops." To construct the space $\Omega'Y$ of Moore loops on Y, we take its elements to be the functions

$$\omega\colon [0,t],\, 0,t \longrightarrow Y,y_0,y_0 \qquad (t \geq 0)\,.$$

We call such a function a "loop of length t." Of course, if $t = 0$, then ω has to be constant. We topologize $\Omega'Y$ so that it is homeomorphic, under the obvious map, to the obvious subset of

$$\Omega Y \times [0,\infty)\,,$$

where $[0,\infty)$ means the half-line $0 \leq t < \infty$. It is then clear that $\Omega'Y$ is equivalent to the ordinary loop-space ΩY. We give $\Omega'Y$ a product map so that the product of a loop of length r and a loop of length s is a loop of length $r+s$; then $\Omega'Y$ becomes a monoid, and ΩY, $\Omega'Y$ become equivalent as H-spaces.

The necessary condition we have just obtained is essentially sufficient: any topological monoid X, such that $\pi_0(X)$ is a group, is equivalent to a loop-space. The proof is not relevant to the present discussion.

The homotopy-theorist is perfectly willing to consider product maps $\mu\colon X\times X \to X$ which have the base-point as a strict unit; for he can use the homotopy extension theorem to deform a product map μ till it has this property. Unfortunately the condition of strict associativity,

$$(xy)z = x(yz)\,,$$

is very inconvenient from the point of view of the homotopy-theorist. He would be quite happy with the condition of homotopy-associativity,

$$\mu(\mu\times 1) \simeq \mu(1\times\mu)\,,$$

but this is not sufficient. Let us see why it is not sufficient.

Let us suppose given a space X, a product map $\mu: X^2 = X \times X \to X$ and a homotopy $h_t: X^3 \to X$ with

$$h_0 = \mu(\mu \times 1), \qquad h_1 = \mu(1 \times \mu).$$

If we write xy for $\mu(x,y)$ this means

$$h_0(x,y,z) = (xy)z, \qquad h_1(x,y,z) = x(yz).$$

Let us now consider maps $X^4 \to X$. We can think of five of them, and they carry the point $(w,x,y,z) \in X^4$ to the compound products displayed in the following diagram.

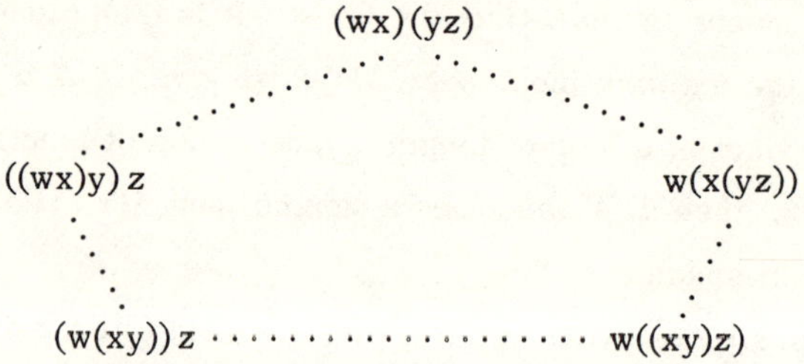

The dotted lines in this diagram represent the five homotopies we can think of between these five maps, as follows.

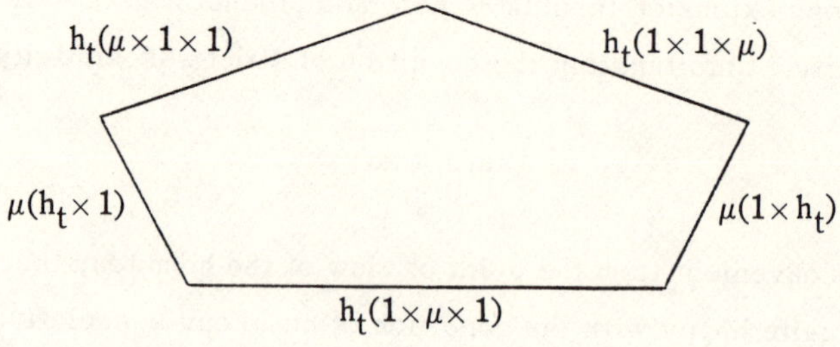

These five homotopies together constitute a map

$$S^1 \times X^4 \longrightarrow X .$$

I can ask whether this map extends to a map (say)

$$H: E^2 \times X^4 \longrightarrow X .$$

Sometimes it does and sometimes it doesn't. Suppose for example that X is a strict topological monoid and μ is its product; then we take

$$h_t(x,y,z) = xyz$$

(independent of t), and our map H does exist – we take

$$H(e,w,x,y,z) = wxyz$$

(independent of $e \in E^2$). But in general a map H does not exist.

We regard the existence of H as a secondary homotopy condition. If the map H does exist, then we can consider maps $X^5 \to X$ and derive a tertiary homotopy condition, and so on.

Of course one has to give a precise meaning to the words "and so on." At this point I explain the general pattern of the work of Stasheff [139]. (For my purposes it is convenient to present the work of Stasheff; but historically much credit is due to the earlier work of Sugawara [144].)

(i) Stasheff defined a sequence of parameter spaces K_m for $m \geq 2$ such that K_m is a cell E^{m-2} with an explicitly described subdivision of its boundary; in particular K_2 is a point, K_3 is a unit interval I and K_4 is the disc E^2 with its boundary subdivided as a pentagon.

(ii) He defined an A_n-space to be a space X provided with maps

$$M_r: K_r \times X^r \longrightarrow X$$

for $2 \leq r \leq n$ satisfying suitable conditions. For example, an A_2-space is a space X provided with a map $M_2: X \times X \to X$; an A_3-space is a space X provided with a map $M_2\colon X\times X \to X$ and an explicit homotopy M_3 between $M_2(M_2\times 1)$ and $M_2(1\times M_2)$; and so on.

(iii) More precisely, the definition is by induction. Suppose that X is an A_{n-1}-space, so that it comes provided with maps $M_2, M_3, \cdots, M_{n-1}$. In terms of these maps Stasheff defines a map from $(\partial K_n)\times X^n$ to X. Then X is an A_n-space if it comes provided with a map

$$M_n\colon K_n \times X^n \longrightarrow X$$

extending this map of $(\partial K^n)\times X^n$.

The maps M_n for $n \geq 4$ are called "higher homotopies."

(iv) A space X is an A_∞-space if it has maps M_n for all $n\geq 2$ which make it an A_n-space for all n.

(v) An A_∞-space is an adequate substitute (usable by homotopy-theorists) for a space with a product which is strictly associative.

More precisely, I have simplified the discussion above by concentrating on the associative condition and neglecting the condition that the base-point should act as a unit; but when Stasheff gave the definition of an A_n-space he included the condition that the base-point should act as a unit in a suitable sense.

One should then have the following result: a space X is equivalent to a loop-space ΩY if and only if X is an A_∞-space and $\pi_0(X)$ is a group. This theorem is implicit in [139], except that he makes an implicit assumption that X is connected; the same remark applies to some of the earlier work used by [139], for example, [54].

More precisely again, one should have a result of the following form: the functor Ω gives an equivalence from the category of (connected) complexes Y (with base-point) to a suitably-defined category of spaces with A_∞-structure.

I should sketch the approach by which one proves such theorems. In the theory of fibre-bundles, it is well known that given a topological group G, we can construct a "universal bundle"

$$G \to EG \to BG$$

with fibre G, total space EG and base BG. According to your assumptions and constructions, the space EG is either contractible, or at least weakly equivalent to a point. A suitable textbook reference is [71], especially pp. 51-57. It is also well understood that the fibering

$$G \to EG \to BG$$

is closely analogous to the fibering

$$\Omega X \to EX \to X$$

mentioned in Chapter 1. (Of course the letter E means different things in the two cases, for in one case it is a functor of the fibre G and in the other case it is a different functor of the base X; the context is usually sufficient to avoid confusion.) In fact one should arrange matters so that the functor B ("classifying space") is essentially inverse to the functor Ω ("loop space"). For this purpose, of course, one has to define classifying spaces and universal bundles for spaces G less good than topological groups. For the case in which G becomes a monoid, see [103, 142] and the most recent reference, [94]. As the assumptions on G become less and less, so the sense in which the "universal bundle" is a fibering tend to become weaker and weaker. In the end one defines the "universal bundle" and "classifying space" when G is an A_∞-space. This is what Stasheff did, and similarly for Sugawara.

If G is just an A_n-space one can still define a part of the "classifying space" which is of some interest. For example, if we take $G = S^1$ (with its usual structure as a topological group) and treat it as an A_n-space, we get the fibering

$$S^1 \longrightarrow S^{2n-1} \xrightarrow{\pi} CP^{n-1} ,$$

and using π as an attaching map we get

$$CP^{n-1} \cup_{\pi} e^{2n} = CP^{n} .$$

§2.3. N-*fold and infinite loop spaces*; E_n *and* E_∞ *spaces*

In this section we will examine the natural continuation of the work in the last section, towards iterated loop-spaces $\Omega^n Y$ with $n \geq 2$.

After §2.2 it is more or less clear that one might go on from Stasheff's work to write conditions for an H-space to be a double loop-space. If $X \simeq \Omega^2 Y$, then certainly the product in X is homotopy-commutative; that is, we have

$$\mu \simeq \mu\tau \colon X^2 \to X ,$$

where

$$\tau(x,y) = (y,x) .$$

However this condition by itself is certainly not sufficient; we want an infinity of higher homotopies, on the same general model as the maps M_n of Stasheff. If we assume $X \simeq \Omega^3 Y$ or $X \simeq \Omega^4 Y$ we bring in further infinities of higher homotopies. Moreover there is some sense and purpose in these higher homotopies, for some of them appear in the construction of homology operations on $H_*(\Omega^n Y; Z_p)$ according to Kudo and Araki [77, 78], Browder [47] and Dyer and Lashof [55]; and these homology operations undoubtedly give sensible information which reflects the n-fold loop structure.

We now face the difficulty that it is quite enough work and nuisance to write down the explicit details about the complexes K_n; to go any further we would certainly like to avoid writing down the explicit details and have a machine which constructs them for us – rather as the method of acyclic models in ordinary homology enables us to avoid writing explicit formulae for the cup-i products. All the "machines" in this chapter are devices for the automatic processing of infinitely many higher homotopies;

this is their essential business, whether or not they hide it behind a sleek exterior.

Unfortunately, as I have hinted in §2.1, the construction and use of machines takes work. Therefore this chapter will be written as an essay in machine appreciation; it is not intended to qualify the reader for a mechanic's certificate. However, I hope it may give the general reader an idea of what goes on, and also serve as an introduction to the study of more technical works to which I shall refer.

We first need to create an ecological niche fit to accommodate (a) the complexes K_n of Stasheff, and (b) any other complexes similarly conceived. The definition with which I will begin is that of a "topological PROP" or "category of operators" in the sense of Boardman and Vogt [39, 40]; it is comparable with that of an "operad" in the sense of J. P. May [92], and I will comment on the difference later.

A topological PROP $\mathcal{P}$ will comprise spaces $P_{a,b}$ indexed by pairs of integers $a,b \geq 0$. When our PROP acts on a space X, we shall be given maps

$$P_{a,b} \times X^b \longrightarrow X^a ;$$

so the spaces $P_{a,b}$ are to be thought of as "parameter spaces" analogous to the complexes K_n of Stasheff, which are parameter spaces $P_{1,n}$ since they parametrize maps from X^n to X.

For example, Kudo and Araki [77, 78] work with spaces X which come provided with structure maps $\theta_m: I^m \times X \times X \to X$; so they use parameter spaces $P_{1,2} = I^m$. Again, Dyer and Lashof [55] work with spaces X which come provided with structure maps $(J^n\Sigma_p) \times X^p \to X$, where Σ_p is the symmetric group on p letters and $J^n\Sigma_p$ is the nth join of Σ_p with itself; so they use parameter spaces $P_{1,p} = J^n\Sigma_p$.

To return to the theory, let us start from a space X; if X is locally compact or if we use compactly-generated topologies, then we can form

$$H_{a,b} = (X^a)^{(X^b)} ;$$

and then we have a map

$$H_{a,b} \times X^b \longrightarrow X^a ;$$

an action of the PROP $\mathcal{P}$ on X is to be a set of continuous maps

$$P_{a,b} \to H_{a,b}$$

commuting with all the structure in sight.

Well, what is this structure? for we have not yet said what the structure of $\mathcal{P}$ should be. First, the sets $H_{a,b}$ form a category; we can compose maps

$$h': X^a \leftarrow X^b \quad \text{and} \quad h'': X^b \leftarrow X^c$$

to get a map

$$h'h'': X^a \leftarrow X^c .$$

So we demand that the sets $P_{a,b}$ should be the Hom-sets of a category; the objects of the category can be either the formal symbols "X^a", $a = 0,1,2,3,\cdots$, or more simply the integers $a = 0,1,2,3,\cdots$; and $P_{a,b}$ is to be the set of maps from the object b to the object a. We also ask that this should be a topological category, in the sense that the composition maps

$$P_{a,b} \times P_{b,c} \longrightarrow P_{a,c}$$

should be continuous.

Secondly, given maps

$$f: X^a \leftarrow X^b \quad \text{and} \quad g: X^c \leftarrow X^d$$

we can form

$$f \times g : X^{a+c} \leftarrow X^{b+d} .$$

So we demand that we are given product maps

$$P_{a,b} \times P_{c,d} \longrightarrow P_{a+c,b+d}$$

with the following properties.

(i) These maps are continuous.

(ii) They are strictly associative.

(iii) The identity map in $P_{0,0}$ acts as a strict unit.

(iv) If 1_a is the identity map in $P_{a,a}$, then

$$1_a \times 1_b = 1_{a+b} .$$

(v) $(f \times g)(h \times k) = (fh \times gk)$ whenever this makes sense.

Thus the $\times$ product is a functor of two variables, given on the objects by addition of integers; we have some sort of "category with products."

Thirdly, the symmetric group Σ_a acts on the space X^a. In fact it acts on the right; if we think of a vector $(x_1, x_2, \cdots, x_a) \in X^a$ as a function

$$X \xleftarrow{x} \{1,2,\cdots,a\}$$

and a permutation as a function

$$\{1,2,\cdots,a\} \xleftarrow{\rho} \{1,2,\cdots,a\} ,$$

then it is clear that the order of composition is

$$X \xleftarrow{x} \{1,2,\cdots,a\} \xleftarrow{\rho} \{1,2,\cdots,a\} ;$$

alternatively, if you look closely at the formula

$$(x_{\rho 1}, x_{\rho 2}, \cdots, x_{\rho a})$$

you see that ρ actually does come on the right of x. Thus the function

$$\rho \longmapsto \rho^{*}: \Sigma_a \longrightarrow H_{a,a}$$

is an antihomomorphism (of monoids). So we demand that there should be given an antihomomorphism of monoids

$$\rho \longmapsto \rho^{*}: \Sigma_a \longrightarrow P_{a,a} .$$

This structure should be related to the $\times$ product as follows.

(i) If $\rho \in \Sigma_a$, $\sigma \in \Sigma_b$ then

$$\rho^{*} \times \sigma^{*} = (\rho \times \sigma)^{*} ,$$

where $\rho \times \sigma \in \Sigma_{a+b}$ is the obvious permutation such that the formula holds in $H_{a+b,a+b}$ for any X.

(ii) If $f \in P_{a,b}$ and $g \in P_{c,d}$, then

$$\rho^{*}(f \times g) = (g \times f)\sigma^{*} ,$$

where ρ is the obvious permutation which you would expect to map $X^a \times X^c$ to $X^c \times X^a$ so as to interchange X^a and X^c, while σ is the similar obvious permutation which you would expect to map $X^b \times X^d$ to $X^d \times X^b$.

Some authors work with a homomorphism

$$\Sigma_a \longrightarrow P_{a,a}$$

of monoids, corresponding to

$$\rho \longmapsto (\rho^{-1})^*: \Sigma_a \longrightarrow H_{a,a} ;$$

the difference is trivial.

In some cases the work to be done with a PROP does not demand the use of permutations, and then it is not necessary to take the functions $\Sigma_a \to P_{a,a}$ as part of the given structure; so one has two variants of the definition, one "with permutations" and one "without permutations." But according to MacLane, [83] p. 97, the name "PROP" comes from "product and permutation category", so a PROP without permutations is really only entitled to be called a PRO.

May's notion of an "operad" [92] is similar to the above, except that he only has parameter spaces $P_{1,a}$; therefore the operations of "composition" and "Cartesian product" do not exist separately in an operad – there is only the combined operation

$$f(g_1 \times g_2 \times \cdots \times g_a)$$

where

$$f \in P_{1,a}, \quad g_i \in P_{1,b_i} .$$

Of course May also states a suitable list of axioms on this one operation.

The motivation I have given really motivates the use of operads rather than PROPs; for example, Stasheff's sequence of spaces K_n is an operad (without permutations), but it is not a PROP. It just seems better to explain about "composition" and "Cartesian product" separately first, before explaining the combined operation

$$f(g_1 \times g_2 \times \cdots \times g_a) .$$

In fact the distinction between an operad and a PROP need not trouble us very much. Given a PROP, we can construct an operad by retaining the spaces $P_{1,b}$ and forgetting the spaces $P_{a,b}$ for $a > 1$.

Given an operad, we can construct a PROP from it by allowing the operad to generate "freely"; that is, if we work without permutations we define

$$P_{a,b} = \bigcup_{b_1+b_2+\cdots+b_a=b} P_{1,b_1} \times P_{1,b_2} \times \cdots \times P_{1,b_a} ;$$

working with permutations we have to replace

$$P_{1,b_1} \times P_{1,b_2} \times \cdots \times P_{1,b_a}$$

by

$$P_{1,b_1} \times P_{1,b_2} \times \cdots \times P_{1,b_a} \times_G \Sigma_b$$

where

$$G = \Sigma_{b_1} \times \Sigma_{b_2} \times \cdots \times \Sigma_{b_a} .$$

If we go from PROP to operad to PROP by these constructions, we don't get the same PROP back; so there are more PROPs than operads; but for present purposes we need those PROPs which are equivalent to operads.

It is also necessary to point out that when we speak of a PROP or an operad acting on a space X, the proper definition may have to pay attention to what is to happen to the base-point.

My next task is to convince you that it is possible (and not even much trouble) to construct some operads which have a chance of acting on spaces. I will do this by exhibiting the "little n-cubes operad." This is an operad which acts on every n-fold loop space; so we consider the case $X = \Omega^n Y$. Then

$$\prod_1^b X = (Y,y_0)^{\left(\bigvee_1^b S^n, s_0\right)} .$$

We will construct a space $P_{1,b}$ in which the points are certain maps

$$S^n, s_0 \longrightarrow \bigvee_1^b S^n, s_0 \,.$$

If we do this, then each map

$$p: S^n, s_0 \longrightarrow \bigvee_1^b S^n, s_0$$

will induce

$$p^*: X \leftarrow X^b \,;$$

so our operad will come complete with a ready-made action on $X = \Omega^n Y$.
The maps

$$p: S^n \to \bigvee_1^b S^n$$

to be considered are as follows.

We identify the sphere S^n with $I^n/\partial I^n$. Each map p of $I^n/\partial I^n$ is to be constant at the base-point outside a set of b non-overlapping little cuboids, with their edges parallel to the axes, as in the figure.

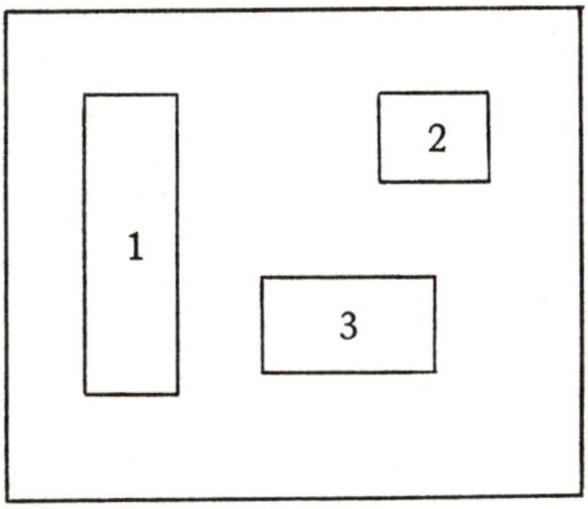

On each such little cuboid the map is to be a linear map

$$(x_1, x_2, \cdots, x_n) \longmapsto (\lambda_1 + \mu_1 x_1, \lambda_2 + \mu_2 x_2, \cdots, \lambda_n + \mu_n x_n)$$

onto one of the summands $S^n = I^n/\partial I^n$ in $\bigvee_1^b S^n$. Each of the b summands in $\bigvee_1^b S^n$ is to be used once and only once; that is, it is to be the image of exactly one little cuboid in $I^n/\partial I^n$.

The figure is drawn for the case $n = 2$, $b = 3$; the little cuboid marked "1" is supposed to be mapped to summand number 1 in $\bigvee_1^3 S^2$, and similarly for the little cuboids marked "2," "3."

The space $P_{1,b}$ is topologized as a subspace of the function space

$$\left(\bigvee_1^b S^n, s_0\right)^{(S^n, s_0)};$$

it is equivalent to say that two maps are close if the corners of corresponding little cuboids are close.

This gadget $\mathcal{P} = \mathcal{P}(n)$ is called the "little n-cubes operad" (or the "little n-cubes PROP," if you make it a PROP). The idea is due to Boardman and Vogt [39]. Clearly this operad is relevant to the geometry which has to be done. It also has the property that the action of the symmetric group Σ_b on $P_{1,b}$ is free.

We may say that a space X is an E_n-space if it comes provided with an action of the little n-cubes operad; or, more loosely, if it comes provided with an action of some other operad equivalent to this one and adapted to whatever technical purposes we have in mind.

The appropriate sort of result is now the following. Roughly speaking, it shows that a space X is an n-fold loop-space if and only if it is an E_n-space; but the result should be more precise.

PRETHEOREM 2.3.1. (i) Ω^n *can be made into a functor, defined on spaces* Y, *and taking values in* E_n*-spaces* X *such that* $\pi_0(X)$ *is a group.*

(ii) *There is a functor* B^n *"inverse to* Ω^n*" from* E_n*-spaces* X *to* (n–1)*-connected spaces* Y.

(iii) *There is a natural transformation*

$$B^n \Omega^n Y \longrightarrow Y$$

which is an equivalence whenever Y *is* (n–1)*-connected.*

(iv) *There is a natural transformation*

$$X \longrightarrow \Omega^n B^n X$$

which is an equivalence whenever $\pi_0(X)$ *is a group.*

If $\pi_0(X)$ is not assumed to be a group, then the relationship between X and $\Omega^n B^n X$ should be such as is discussed in §3.2. It is essentially sufficient to consider this point for the case $n = 1$.

I pause to explain the word "pretheorem," for if I don't the reader may think it is a sort of theorem which has not yet been proved properly. This is not what I mean; I intend no slur on other people's proofs, only on my own statements. As I shall use the word, a pretheorem is a statement sufficient (I hope) to explain the general drift and intention of a result, but not necessarily sufficient to pin down all the technical details which need to be pinned down before you can have a theorem. The word "pretheorem" should usually encourage the reader to refer to the original sources. It should mean that there exists at least one set of technical details supporting a theorem of the sort sketched; but it usually means that the literature offers a choice of such theorems. In that case you should shop around and see if one or other of the competing authors offers a specific choice of technical details which is particularly convenient for your purpose. Or if you cannot maximize convenience, you try to minimize inconvenience.

In the case of (2.3.1), precise theorems of this form may be found in [38] p. 57 and [92] pp. 3, 128.

All comment on the methods used to prove such theorems is postponed to the next section.

We now want to pass to a limit as $n \to \infty$. It is certainly possible to embed the operad $\mathcal{P}(n)$ of little n-cubes in the operad $\mathcal{P}(n+1)$ of little (n+1)-cubes; given any map $S^n \to \bigvee_1^b S^n$, we take its product with the identity map of the x_{n+1}-coordinate. Let $\mathcal{P}(\infty)$, the "little-cubes operad," be $\bigcup_n \mathcal{P}(n)$, the union over n of the little n-cubes operads. It is possible to arrange the functor Ω^∞ from spectra to spaces so that this operad $\mathcal{P}(\infty)$ acts on the spaces $\Omega^\infty Y$; this is essentially the same point as that involved in rearranging spectra so as to get homeomorphisms

$$X_n \cong \Omega X_{n+1}$$

(see §1.4); for details the reader is referred to May [89]. We may say that a space X is an E_∞-space if it comes provided with an action of this operad $\mathcal{P}(\infty)$; or, more loosely, if it comes provided with an action of some other operad equivalent to this one and adapted to whatever technical purposes we have in mind.

The principal good property of this operad $\mathcal{P}(\infty)$ is as follows. For the operad $\mathcal{P}(n)$ of little n-cubes, each space $P_{1,b}(n)$ is (n–2)-connected; it follows that for the operad $\mathcal{P}(\infty)$, each space $P_{1,b}(\infty)$ is contractible.

Boardman and Vogt call X an "E-space" if they are given an action on X of a PROP $\mathcal{P}$ (with permutations) such that the space $P_{1,b}$ is contractible for each b. This is essentially the same notion as that of "E_∞-space" given above. The idea is that the contractibility of $P_{1,b}$ guarantees the existence of all the higher homotopies one can ever need. For example, there is a point μ in $P_{1,2}$, which will act as a product map μ for X and make X an H-space. Let τ be the non-trivial element in Σ_2, so that τ acts on X^2 by the usual switch map $\tau(x,y) = (y,x)$; then the points μ and $\mu\tau$ in $P_{1,2}$ must be joined by a path, and so X is a homotopy-commutative H-space; the points $\mu(\mu\times 1)$ and $\mu(1\times\mu)$ in

$P_{1,3}$ must be joined by a path, so X is a homotopy-associative H-space; similarly X is an A_∞-space, and so on. For a given n, some of these constructions can be carried out inside the operad $\mathcal{P}(n)$ of little n-cubes and some cannot. Boardman and Vogt originally proposed the term "homotopy-everything H-space" for an E-space; however, they later withdrew it, for reasons which need not delay us.

The following statement is analogous to (2.3.1). Roughly speaking, it shows that a space X is an infinite loop space if and only if it is an E_∞-space; but the result should be more precise.

PRETHEOREM 2.3.2. (i) Ω^∞ *can be made into a functor, defined on spectra* Y, *and taking values in* E_∞*-spaces* X *such that* $\pi_0(X)$ *is a group.*

(ii) *There is a functor on* B^∞ *"inverse to* Ω^∞*" from* E_∞*-spaces to connective spectra.*

(iii) *There is a natural transformation*

$$B^\infty \Omega^\infty Y \longrightarrow Y$$

which is an equivalence whenever Y *is connective.*

(iv) *There is a natural transformation*

$$X \longrightarrow \Omega^\infty B^\infty X$$

which is an equivalence whenever $\pi_0(X)$ *is a group.*

If $\pi_0(X)$ is not assumed to be a group, then the relationship between X and $\Omega^\infty B^\infty X$ should be such as is discussed in §3.2.

Precise theorems of this form may be found in [16, 39, 40, 92, 93, 127].

All comment on the methods used to prove such theorems is postponed to the next section.

The extent to which (2.3.1) and (2.3.2) fail to pin down details is easily exposed. For example, (2.3.1)(i) claims that Ω^n can be made into a functor; but this is only meaningful if the E_n-spaces have been made into a category. What are the morphisms? Clearly I didn't tell you. What should be the morphisms? One candidate for a definition is obvious. Suppose that X and Y admit actions from the same operad $\mathcal{P}$; then we can consider maps $f: X \to Y$ which commute (strictly) with all the structure maps; that is, the diagram

$$\begin{array}{ccc} X & \xleftarrow{\;p\;} & X^b \\ {\scriptstyle f}\downarrow & & \downarrow{\scriptstyle f^b} \\ Y & \xleftarrow{\;p\;} & Y^b \end{array}$$

commutes (strictly) for all $p \in P_{1,b}$. This definition has the virtue of simplicity; it provides the easiest way to put precise sense into the statement; and it is sufficient for the applications of current interest. However, it is not the only possible definition. For example, when one defines the notion "map of H-spaces," one asks that the diagram

$$\begin{array}{ccc} X & \xleftarrow{\;\mu\;} & X^2 \\ {\scriptstyle f}\downarrow & & \downarrow{\scriptstyle f^2} \\ Y & \xleftarrow{\;\mu\;} & Y^2 \end{array}$$

commutes up to homotopy – not strictly. Starting from such an example, one might arrive at other definitions which, from our present point of view, have the virtue of homotopy-invariance and the defect of elaborate complexity. Such an approach may be found in [40] and [53] Part V.

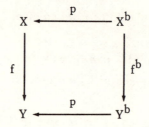

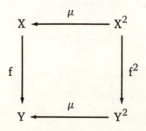

Next I should point out that results such as (2.3.1) and (2.3.2) do not exhaust the business of the theory. It was clearly necessary to explain them first, because of their importance to our general theme; in the words of J. P. May, they constitute a "recognition principle" by which we can recognize whether a given space is an n-fold loop-space, or an infinite loop space. But the theory has other business too.

First, recall from Chapter 1 that I. M. James [73] gave a "model" for ΩS^n which allowed one to understand the structure of this space. In fact he gave a model for $\Omega\Sigma X$, where X is any connected space; and from this model one can read off the results on $H_*(\Omega(S^{n_1} \vee S^{n_2} \vee \cdots \vee S^{n_d})$ which were mentioned in Chapter 1. Later, Milgram [102] gave a model for $\Omega^n\Sigma^n X$ in the same spirit. The theory should provide and does provide such models for $\Omega^n\Sigma^n X$ and $\Omega^\infty\Sigma^\infty X$; see [92]. May calls this the "approximation theorem." The theory should explain and does explain in what sense the spaces $\Omega^n\Sigma^n X$, $\Omega^\infty\Sigma^\infty X$ (or their models) are "free" objects in the category of E_n-spaces or E_∞-spaces. The proofs of the "approximation theorem" and the "recognition principle" should be made to support one another.

There is a clear profit in having control over spaces such as $\Omega^n\Sigma^n X$. For example, Snaith has given a stable splitting of $\Omega^n\Sigma^n X$ [133] and this has been used by Mahowald to construct an interesting new family of elements in the 2-primary stable homotopy of spheres. The interest thus aroused has led to new proofs of Snaith's result by Cohen and Taylor, and to generalizations of it by May, but I do not have references yet.

Again, to the eye of a topologist whose tastes are algebraic, an "E_n-structure" or an "E_∞-structure" is a rather flabby collection of higher homotopies; one longs to squeeze the water out of it and see what actual invariants there are in it. Certainly there are many; and to pass from the given geometry to well-chosen invariants which can be used in calculation is part of the business of the theory. In this direction, see [90, 91, 152].

Finally, the theory must seek its validation by applications to concrete problems and special cases of interest.

§2.4. *Methods*

In this section I will say something about the methods which may be used to prove results of the sort stated in §2.3. My object is still to avoid detail, but to try to transmit a few ideas which may help to enlighten the reader.

The form of (2.3.1), (2.3.2) shows that the first task is to set up the functors B^n, B^∞. After that, the second task is to establish their properties. And in order to set up the functors, there are two basic strategies. Boardman and Vogt in [40] set up a one-step functor B and proceed to iterate it; May in [92] sets up the functor B^n at one blow.

To be slightly more precise, Boardman and Vogt suppose given an E_∞-space X. They then replace X by an equivalent monoid $Y = MX$. As I shall explain below, Boardman and Vogt only use structures which are homotopy-invariant; so they get as much structure on Y as they had on X. They then take a classifying space BY. Next they place an E_∞-structure on BY; since BY is constructed out of $Y, Y^2, Y^3, \cdots$ and constant auxiliary spaces (such as cells), it is more or less plausible that you can construct one structure-map on BY by using an infinite amount of data in the form of maps and homotopies from Y^b to Y^a for various values of a and b. Once BY is an E_∞-space, the construction can be iterated.

I will comment on May's construction of B^n later.

The differences of technique which one may observe between these authors are perhaps correlated with differences of philosophy and outlook. I trust that I can show my sympathy for both sides. It seems that Boardman and Vogt feel that there really should be a theory of homotopy-invariant structures, and that someone should take the trouble to set this theory up properly for its own sake. For example, one sets up the theory of H-spaces so that if X is an H-space, then any space Y equivalent to

X is also an H-space. Therefore one should set up the theory of A_n-spaces or E_∞-spaces to have the corresponding property: if X is a space with an A_n-structure or an E_∞-structure, and if $f: X \to Y$ is a homotopy equivalence, then there is one and (essentially) only one way to put an A_n-structure or an E_∞-structure on Y so that f becomes an equivalence of A_n-spaces or E_∞-spaces.

In order to carry out this program, it seemed reasonable to Boardman and Vogt to use PROP's which are "free" in a certain sense; the precise sense need not delay us here. Boardman and Vogt construct PROP's which are "free" in this sense. The construction involves combinatorial machinery (trees) [37, 38]. Essentially this machinery corresponds to the "grammar" of "words" in the letters $M_2: K_2 \times X^2 \to X$, $M_3: K_3 \times X^3 \to X$, or similar letters for similar operations. It is this combinatorial machinery which gives the work of Boardman and Vogt much of its individual flavor.

By contrast, May feels (it seems) that the task of the theory is to prove the theorems as quickly as possible, so that we can all go back to our proper business which we enjoy so much, namely computing things which do have invariant meaning, such as homology operations. And to this end he makes adroit use of the following device. If you want to compare X and Y, don't try to construct a map $f: X \to Y$ and don't try to construct a map $g: Y \to X$; construct instead a new object Z which maps to both X and Y.

May's construction of the functor B^n is by a variant of the bar construction; I recall that the original bar construction was introduced by Eilenberg and MacLane [57], and serves to compute the homology of classifying-spaces. I must next explain the ideas behind its generalization. In order to make a generalized bar construction, you need a particular sort of functor $T: C \to C$ from a category C to itself. We can see what is going on from an example.

EXAMPLE 2.4.1. Suppose given an algebra A over some ring of coefficients R; we want to perform the bar construction on A in the usual sense. We take C to be the category of R-modules and T to be the functor

$$T(M) = A \otimes_R M .$$

The axioms which we put on our functor T are obvious from this example. We suppose given a natural transformation

$$\mu : T^2 \to T ,$$

which in the example becomes the map

$$A \otimes_R (A \otimes_R M) \longrightarrow A \otimes_R M$$

induced by the product map

$$A \otimes_R A \to A .$$

We suppose given a natural transformation

$$\eta : 1 \to T ,$$

which in the example becomes the map

$$M = R \otimes_R M \longrightarrow A \otimes_R M$$

induced by the unit map

$$R \to A .$$

Moreover, the given maps μ, η should make the following diagrams commutative.

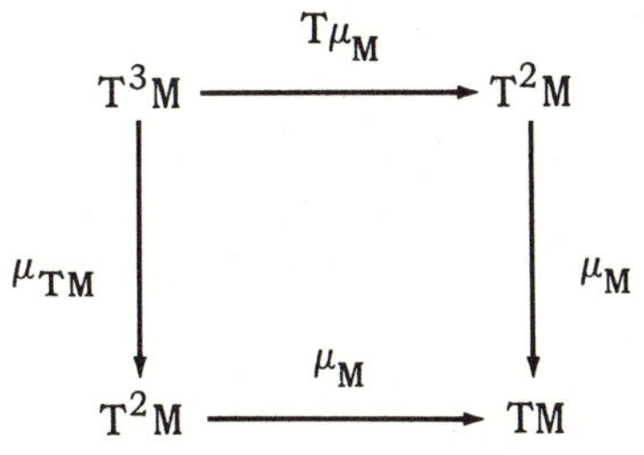

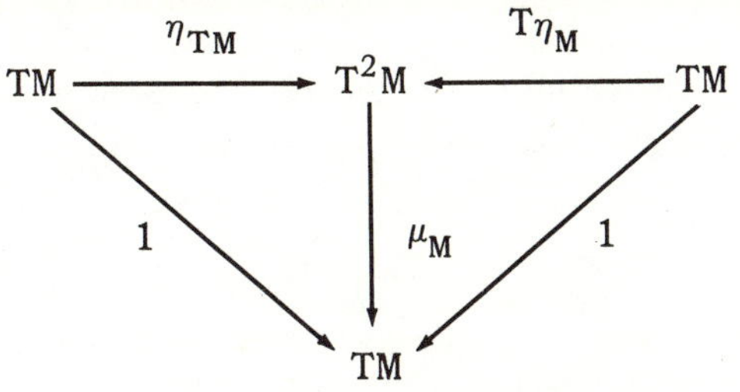

Such a functor T equipped with μ and η is analogous to an algebra in homological algebra; I would like to call it an algebra-functor; unfortunately the categorists have already given it a name, and it is called a "monad" (or "triple"). See [84] p. 133.

EXAMPLE 2.4.2. Let C be the category of compactly-generated spaces with base-point; let TX be $\Omega^n\Sigma^nX$. We certainly have a natural transformation

$$\eta: X \longrightarrow \Omega^n\Sigma^nX$$

(adjoint to $1: \Sigma^nX \to \Sigma^nX$), and another $\Sigma^n\Omega^nY \to Y$, from which we derive

$$\mu: \Omega^n\Sigma^n\Omega^n\Sigma^nX \longrightarrow \Omega^n\Sigma^nX .$$

The study of this example by J. Beck [28] is acknowledged by May as his source for the idea.

EXAMPLE 2.4.3. More generally, any time we have a pair of adjoint functors we can obtain a monad by the same steps displayed in (2.4.2) for the special case of Ω^n and Σ^n.

Let us return to Example 2.4.1, in which C is "R-modules" and T(M) is $A \otimes_R M$. How would we describe an A-module? It's an object M of C provided with a map

$$\nu : A \otimes_R M \to M .$$

We can make the same definition in general. Suppose given a category C and a monad $T: C \to C$; then a "T-object" is an object $M \in C$ provided with a map

$$\nu : TM \to M$$

such that the following diagrams are commutative.

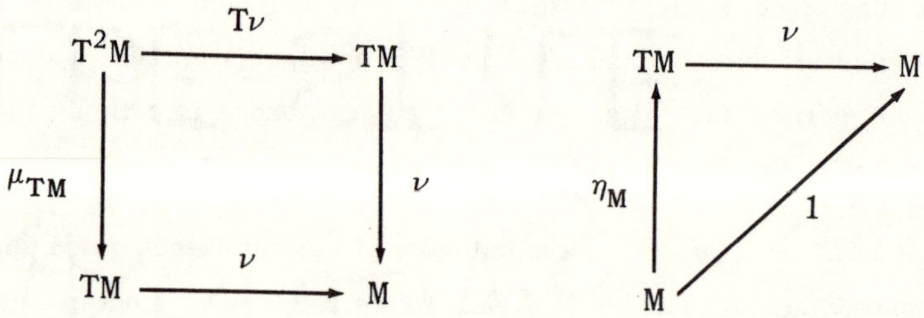

Category-theorists seem to be stuck with the term "T-algebra" for a T-object; in view of the analogies which are important to us here, "T-module" would be more appropriate.

EXAMPLE 2.4.4. Let C be the category of compactly-generated spaces with base-point; let TX be $\Omega^n\Sigma^n X$, as in (2.4.2); then any object $\Omega^n Y$ is a T-object.

Let us return again to Example 2.4.1. The classical bar construction may be used to compute $\mathrm{Tor}_A(L,M)$. We have provided a niche for the algebra A and a niche for the module M, and we are ready to construct

$A \otimes_R A \otimes_R \cdots \otimes_R A \otimes_R M$, but we have not yet provided a niche for the module L and we are not ready to construct $L \otimes_R A \otimes_R \cdots \otimes_R A \otimes_R M$. Evidently we must legislate for one more functor from C to C, and that is the functor $S(N) = L \otimes_R N$.

In general, then, we suppose given a functor

$$S : C \to C'.$$

(It might take values in a new category C'.) We suppose given a natural transformation

$$\lambda : ST \to S;$$

in our example this becomes the map

$$L \otimes_R A \otimes_R N \longrightarrow L \otimes_R N$$

induced by the action map

$$L \otimes_R A \to L .$$

We suppose that S and λ satisfy the obvious axioms, that is, they make the following diagrams commute.

$$\begin{array}{ccc} ST^2N & \xrightarrow{S\mu_N} & STN \\ {\scriptstyle \lambda_{TN}}\downarrow & & \downarrow{\scriptstyle \lambda_N} \\ STN & \xrightarrow{\lambda_N} & SN \end{array} \qquad \begin{array}{ccc} STN & \xrightarrow{\lambda_N} & SN \\ {\scriptstyle S\eta_N}\uparrow & \nearrow{\scriptstyle 1} & \\ SN & & \end{array}$$

Such a functor may be called a (right) "T-functor" (or right module functor over the algebra-functor T).

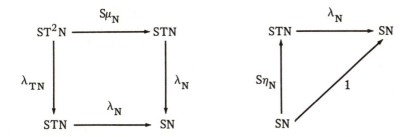

EXAMPLE 2.4.5. Σ^n is a right module functor over the algebra-functor $\Omega^n\Sigma^n$.

At this point the reader may well see a further possibility. Let us define also left module functors over the algebra-functor T; then Ω^n will be a left module functor over $\Omega^n\Sigma^n$, and a "T-object" will be a left module functor from the one-point category to C. Then we shall be able to carry category-theory to its logical conclusion; for not only will we do without elements inside our objects, we will avoid mentioning objects inside our categories; everything in sight will be a functor.

Rest easy; you can trust the categorists not to miss a point like that.

A basic observation of May is now as follows. Let C be the category of compactly generated spaces with base-point, and let $\{P_{1,n}\}$ be an operad. Then we can convert $\{P_{1,n}\}$ into a monad, that is, a functor $P: C \to C$ of the sort described above, so that an action of the operad $\{P_{1,n}\}$ on X is precisely equivalent to giving a structure map $PX \to X$ making X a P-object. The construction is basically obvious; to construct X, you start from

$$\coprod_n P_{1,n} \times X^n$$

and impose a suitable equivalence relation; see [92] p. 13.

The idea is now that the monad P can serve as an adequate "tame" substitute for the "wild" monad $\Omega^n\Sigma^n$. Here "tame" implies that P is usable because it is under our control.

THEOREM 2.4.6. *If* $\{P_{1,n}\}$ *is a suitable operad, equivalent to the little n-cubes operad, and if* X *is connected, then* PX *is weakly equivalent to* $\Omega^n\Sigma^n X$.

See [92] pp. 15, 50. This is the "approximation theorem" to which I referred at the end of §2.3. In other words, PX will serve as a "model" for $\Omega^n\Sigma^n X$.

In order to proceed further with the bar construction, we need to talk about simplicial methods. Let σ^n be the standard n-simplex in R^{n+1}, given by $x_0 + x_1 + \cdots + x_n = 1$, $x_i \geq 0$ for each i; its vertex v_i is the point with $x_i = 1$, $x_j = 0$ for $i \neq j$. We now define a category Δ in which the objects are the standard simplexes $\sigma^0, \sigma^1, \sigma^2, \cdots$; or if you like to replace the geometry by combinatorics, you may replace each simplex by its vertices and say that the objects of the category are the finite sets $\{0\}, \{0,1\}, \{0,1,2\} \cdots$. With the latter interpretation, the morphisms from $\{0,1,\cdots,n\}$ to $\{0,1,\cdots,m\}$ are the functions $f\colon \{0,1,\cdots,n\} \longrightarrow \{0,1,\cdots,m\}$ such that $i \leq j$ implies $fi \leq fj$; so the morphisms are the non-decreasing functions. Such a function corresponds in the geometrical interpretation to the simplicial map from σ^n to σ^m which carries each vertex v_i to v_{fi}.

We now have a standard definition: a "simplicial set" is a contravariant functor from Δ to sets. For example, if X is a topological space, we can construct its total singular complex; this is a simplicial set K, which attaches to each $n = 0,1,2,\cdots$ the set K_n of continuous functions $f\colon X \leftarrow \sigma^n$; and if $g\colon \sigma^n \leftarrow \sigma^m$ is a morphism in Δ, then $g^*\colon K_n \to K_m$ is defined by taking $g^*(f)$ to be the composite

$$X \xleftarrow{\;f\;} \sigma^n \xleftarrow{\;g\;} \sigma^m .$$

Sometimes I will adopt a similar notation for a general simplicial set K, and write fg instead of $g^*(f)$; here of course $f \in K_n$ and $g\colon \sigma^n \leftarrow \sigma^m$ is a map in Δ.

When we give a simplicial set K, it is of course sufficient to give g^* as g runs over a set of generators for the morphisms of Δ. (Generation is by composition.) There is a unique minimal set of generators, and it consists of the following maps.

(i) The maps $d_i\colon \{0,1,\cdots,n-1\} \longrightarrow \{0,1,\cdots,n\}$, where d_i omits the value i and takes the other values once each. These maps are called face maps; more generally, the non-decreasing injections are called face maps.

(ii) The maps $s_i: \{0,1,\cdots,n+1\} \longrightarrow \{0,1,\cdots,n\}$, where s_i takes the value i twice and takes the other values once each. These maps are called degeneracy maps; more generally, the non-decreasing surjections are called degeneracy maps.

These generators satisfy relations which it is usually unnecessary to remember; in theoretical work we can talk about the category Δ and leave it to look after its own structure. (Compare the treatment of the singular complex given above.)

For a textbook reference on simplicial sets, see [88].

Topologists generally feel that a simplicial set is an acceptable combinatorial substitute for a topological space. In fact the "singular complex" described above provides a functor from topological spaces to simplicial sets; there is also a functor in the reverse direction, which is called "geometrical realization"; I describe this next.

Given a simplicial set K, form

$$\coprod_n K_n \times \sigma^n ;$$

then identify (xg,y) with (x,gy) for each $x \in K_n$, $g: \sigma^n \leftarrow \sigma^m$ and $y \in \sigma^m$. The result is written $|K|$.

There are other simplicial notions besides the notion of a simplicial set. They follow a standard pattern. For example, a "simplicial group" is a functor from Δ to groups; a "simplicial ring" would be a functor from Δ to rings; and a "simplicial space" is a functor from Δ to topological spaces.

The geometrical realization $|K|$ makes sense also when K is a simplicial space. In the account above, we have to take $K_n \times \sigma^n$ as the product of two topological spaces (where previously we used the discrete topology on K_n). We then take the disjoint union

$$\coprod_n K_n \times \sigma^n$$

and pass to the quotient space as before.

We can now give the following rough summary of the generalized bar construction of [92] §9 et seq. Let P be a monad defined on spaces; let X be a P-space; and let S be a P-functor from spaces to spaces (that is, a right module-functor over the algebra-functor P). Then the sequence of spaces

$$K_n = SP_nX$$

(provided with suitable maps) is a simplicial space. (For this purpose one does need to use the detailed structure of Δ, on which I poured scorn a page or two back.) We define a generalized bar construction by

$$\text{Bar}\,(S,P,X) = |K|$$

where $K_n = SP^nX$. (In the literature it is usually written $B(S,P,X)$, but I want to avoid confusion with the use of B for a classifying space.)

May's construction of his functor B^n is now as follows. Take a suitable operad $\{P_{1,n}\}$ equivalent to the operad of little n-cubes. Convert it into a monad P. Then for any P-space we can form

$$B^nX = \text{Bar}\,(\Sigma^n,P,X)$$

([92] p. 128).

This completes my remarks on May's methods.

§2.5. *Segal's machine*

In this section I will describe some machinery which appears somewhat different from that surveyed in §2.2 - §2.4, but is intended for a similar purpose. This machine is due to G. Segal [125, 127]. The first account to appear in print was by D. W. Anderson [16].

I will begin by trying to point out two basic ideas in Segal's approach, and the first idea is as follows. In the approach of Boardman-Vogt and May (§2.2 - §2.4) part of the structure we put on X was a map

$$P_{1,n} \times X^n \longrightarrow X .$$

Here we replaced X^n by the bigger space $P_{1,n} \times X^n$ in order that the maps $\mu(\mu \times 1): X^3 \to X$ and $\mu(1 \times \mu): X^3 \to X$ might be different, but yet connected by a homotopy. Topologists refer to the extra bulk of $P_{1,n} \times X^n$ as "connective tissue" (or more shortly, flab). Segal's first idea is that for this purpose there is no need to insist that the bigger space actually comes as the Cartesian product of a parameter space $P_{1,n}$ and n copies of X; it can be bigger and flabbier yet, provided that it has the right algebraic properties and the right homotopy type.

Segal's second idea goes towards formulating the required algebraic properties. In brief, it says that category-theory and simplicial sets provide a slick way of keeping track of formulae like $\mu(\mu \times 1)$ and $\mu(1 \times \mu)$.

Let me take the latter point first. To begin with I claim that the simplicial category Δ introduced in §2.4 may be identified with a set of formulas like $\mu(\mu \times 1)$ which (i) make sense for all topological monoids M, and (ii) suffice to construct the classifying space BM.

More precisely, suppose given a topological monoid M. Then I must explain that one of the constructions for BM begins by constructing a certain simplicial space K. The elements of K_n are defined to be the simplicial 1-cocycles of σ^n with values in M. (This is the same construction that Eilenberg and MacLane used when M is a discrete group G to construct a simplicial set of type $(G,1)$.) Any map $f: \sigma^n \to \sigma^m$ in the category Δ defines an induced map of cocycles, that is a map $f^*: K_n \leftarrow K_m$. If there is any doubt what "1-cocycles" are, one can give the following categorical definition. We regard the monoid M as a category with one object, so that the morphisms of the category are the elements of M and composition is by multiplication in M. We regard the object σ^n as the category

$$0 \leftarrow 1 \leftarrow 2 \leftarrow 3 \leftarrow \cdots \leftarrow n ,$$

so that its objects are the integers $0,1,2,\cdots,n$ and there is a morphism from i to j whenever $i \geq j$. Then a "1-cocycle on σ^n" is a covariant functor from the category σ^n to the category M. A morphism $f: \sigma^n \to \sigma^m$ in the category Δ may be regarded as a functor from

$$0 \leftarrow 1 \leftarrow 2 \leftarrow \cdots \leftarrow n$$

to

$$0 \leftarrow 1 \leftarrow 2 \leftarrow \cdots \leftarrow m ,$$

so we get induced maps as required. In any case, we get a good simplicial space K.

In this construction we have

$$K_n \cong M^n = M \times M \times \cdots \times M \quad (n \text{ factors}) .$$

In fact, the (1-1)-correspondence sends a cocycle u to its values

$$u(v_0 v_1), \ u(v_1 v_2), \ \cdots, \ u(v_{n-1} v_n) ,$$

or equivalently, it sends a functor u to its values on the morphisms

$$0 \leftarrow 1, \quad 1 \leftarrow 2, \quad \cdots, \quad (n-1) \leftarrow n .$$

Therefore, the category Δ may be regarded as a category of formulas for maps $M^n \leftarrow M^m$, formulas which make sense for all monoids. And this class of formulas is sufficient to construct the classifying space of M, because one construction for the classifying space of M is

$$BM = |K|$$

(for this "geometrical realization," see §2.4).

This "class of formulas" does contain the product map $\mu: M^2 \to M$ (which corresponds to the map $d_1: \sigma^1 \to \sigma^2$). However, it is unnecessary to worry in detail how it contains the maps $\mu(\mu \times 1)$, $\mu(1 \times \mu)$ and the information that they are equal; the moral is that you don't need to write down formulae like $\mu(\mu \times 1)$ and $\mu(1 \times \mu)$, because the category Δ will serve as a storehouse for all such formulae. And this has an extra advantage, because other people have invested a lot of work in building up the theory of simplicial sets.

We may now express Segal's idea as follows, so far as it applies to single loop-spaces. Let X be a homotopy-associative H-space; then we can regard X as giving a functor

$$\Delta \xrightarrow{X} \text{(homotopy category)}$$

from the simplicial category Δ to the category of spaces and homotopy classes of maps; that is, the functor sends σ^n to $X^n = X \times X \times \cdots \times X$, d_1 to μ and so on. We shall be able to construct a classifying space for X if we can lift this functor to a functor

$$\Delta \xrightarrow{K} \text{(spaces)},$$

for then we can form $|K|$. Here the word "lift" should mean that the diagram

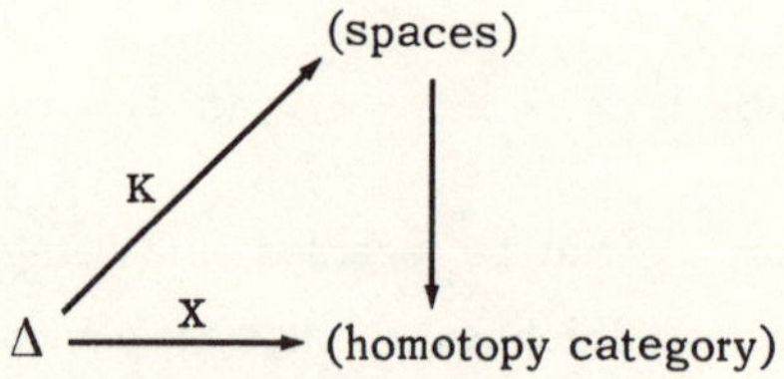

should commute up to natural equivalence. It remains to spell out this last requirement.

Let $i_1, i_2, \cdots, i_n : \sigma^1 \to \sigma^n$ be the maps in Δ which embed σ^1 as the edges $v_0 v_1$, $v_1 v_2$, $\cdots$, $v_{n-1} v_n$ of σ^n. If K is a simplicial space, we can form a map

$$K_n \longrightarrow K_1 \times K_1 \times \cdots \times K_1 \quad (n \text{ factors})$$

with components $i_1^*, i_2^*, \cdots, i_n^*$. Then Segal says that a simplicial space K is "special" if this map

$$K_n \longrightarrow K_1 \times K_1 \times \cdots \times K_1$$

is a homotopy equivalence for each n. Since Segal uses the term "Δ-space" for a simplicial space, he actually talks about "special Δ-spaces."

So the analogue which Segal proposes for an A_∞-space X in the sense of Stasheff is a special Δ-space K with $K_1 = X$. In this case K_n is homotopy-equivalent to $X^n = X \times X \times \cdots \times X$, and K_n plays the same role as the product $P_{1n} \times X^n$ in §2.2-§2.4.

It is understandable that all this should work perfectly well, and so we can move on to see how the same ideas apply to infinite loop spaces. The first step is to replace Δ by a category Γ of formulae such as μ, $\mu\tau, \cdots$, which may involve permutations, and make sense on commutative monoids.

The objects of Γ are taken to be the finite sets. It will do no harm if the reader thinks of the sets $\{1,2,\cdots,n\}$ for $n = 0,1,2,\cdots$, but it is technically convenient to use all finite sets. For each such set σ, let $P(\sigma)$ be the set of subsets of σ. A *morphism* from σ to τ in Γ is to be a function $\theta : P(\sigma) \to P(\tau)$ which preserves disjoint unions. (In particular it is determined by giving its values on the 1-element subsets $\{i\} \subset \sigma$; and these values must be disjoint subsets of τ.)

Just as we identified Δ (or more accurately, its opposite) with a category of formulae which made sense on monoids, so we can identify the opposite of Γ with a category of formulae which make sense on commutative monoids. For this purpose, let A be an abelian monoid, and let

$$\theta : P(\sigma) \to P(\tau)$$

be a map; we want to define a map

$$\theta^*: \mathop{\mathsf{X}}_{i\epsilon\sigma} A \longleftarrow \mathop{\mathsf{X}}_{j\epsilon\tau} A .$$

If $a \epsilon \mathop{\mathsf{X}}_{j\epsilon\tau} A$, we define $\theta^* a$ by giving its ith component as

$$(\theta^* a)_i = \prod a_j \mid j \epsilon \theta\{i\} ,$$

where the product Πa_j is formed in the abelian monoid A.

Obviously a formula which makes sense on all monoids makes sense on commutative monoids. This defines an embedding $\Delta \to \Gamma$. More precisely, we send the object σ^n to the finite set $\{1,2,\cdots,n\}$; and if f is a morphism in Δ, regarded as a non-decreasing map

$$f: \{0,1,2,\cdots,n\} \longrightarrow \{0,1,2,\cdots,m\}$$

then we define a morphism $f': P\{1,2,\cdots,n\} \longrightarrow P\{1,2,\cdots,m\}$ in Γ by

$$f'\{i\} = \{j \mid f(i-1) < j \le f(i)\} .$$

A "Γ-space" K is a contravariant functor from Γ to spaces; it follows from the above that a Γ-space yields a Δ-space. We say that a Γ-space is "special" if the corresponding Δ-space is special in the sense defined above.

The concept of a "special Γ-space" is the one which Segal offers as a substitute for the Boardman-Vogt-May notion of an "E_∞-space." From what we have said so far it seems likely to work perfectly well; let us just confirm this by looking at the notion of "classifying space" B. We have seen that this circle of ideas makes it convenient to deliver a classifying space in the form of a single space, but now we want an outcome which is a Γ-space, that is, a functor from Γ to spaces. So let K be the Γ-space which we want to classify, and σ be an object of Γ. We define the functor BK on σ to be the geometric realization of the functor

$$\tau \mapsto K(\sigma \times \tau) .$$

Of course it has to be checked that this works, but clearly it is quite a slick construction.

This approach was originally intended to give a theory for infinite loop spaces only and not for n-fold loop spaces, but that covers the most interesting applications. Moreover, a variant of Segal's machine, intended for use on n-fold loop-spaces, has been sketched by Cobb [52].

Another machine conceived in a simplicial spirit is due to Barratt and Eccles [22, 23, 24, 25, 26].

§2.6. *Categorical input*

The question now arises: in practice, where do you get all the higher homotopies to feed into a machine such as those considered in §2.2 - §2.5? The answer is that you get them from some context where you have products which "in principle" are associative and commutative. For example, in the case of the bundle-theories considered in §1.8, it is plausible that all the Whitney sums are "in principle" associative and commutative; and in [40, 99] some care is put into justifying this intuition. The technical question is, how do you attach rigorous meaning to those words "in principle"? In this section we will study one proposed answer.

It seems best if we begin by looking at two concrete examples.

EXAMPLE 2.6.1. C is the category in which the objects are the sets $\{1,2,3,\cdots,n\}$ (for $n \geq 0$); there are no morphisms from the nth object to the mth object unless $n = m$; then the morphisms are the (1-1) correspondences, i.e. they form the symmetric group Σ_n.

EXAMPLE 2.6.2. We are given a (discrete) commutative ring R, and C is the category in which the nth object is the free module R^n. There are no morphisms from the nth object to the mth object unless $n = m$; then the morphisms are the R-linear (1-1) correspondences, i.e. they form the matrix group $GL(n,R)$.

It is a common feature of these two categories that one has a bifunctor $\square : C \times C \to C$ which is strictly associative and has a strict unit. More precisely, in both cases we define $\square$ on objects by

$$n \square m = n + m ;$$

certainly this is strictly associative and has 0 as a strict unit. We must also define $\square$ on maps. In (2.6.2), suppose given $A \in GL(n,R)$ and $B \in GL(m,R)$; we define

$$A \square B = \begin{bmatrix} A & 0 \\ 0 & B \end{bmatrix} .$$

Clearly this is strictly associative and has the unique element of $GL(0,R)$ as a strict unit. Similarly, in (2.6.1), suppose given $\sigma \in \Sigma_n$ and $\tau \in \Sigma_m$; we define $\sigma \square \tau \in \Sigma_{n+m}$ so that it acts like σ on $\{1,2,\cdots,n\}$ and like τ on $\{n+1, n+2, \cdots, n+m\}$. Clearly this is strictly associative and has the unique element of Σ_0 as a strict unit.

Categorists express the properties we have checked in these examples by saying that C is a *strict monoidal category*. (See [84] pp. 157-158; however, some readers may prefer [85], a reference whose brevity is almost sufficient recommendation.)

In examples 2.6.1 and 2.6.2, although the product $\square$ is "in principle" commutative it is not strictly so; for example,

$$\begin{bmatrix} A & 0 \\ 0 & B \end{bmatrix} \neq \begin{bmatrix} B & 0 \\ 0 & A \end{bmatrix}$$

in general.

More generally, suppose given a category C equipped with a product $\square$, as above; categorists say that C is *symmetric* if it comes equipped with a natural isomorphism

$$c : \square \cong \square\tau : C \times C \to C ,$$

where

$$\tau : C \times C \to C \times C$$

is the switch map, defined by

$$\tau(X,Y) = (Y,X)$$

$$\tau(f,g) = (g,f) .$$

This natural isomorphism is required to satisfy suitable axioms; see [84] p. 180. The axioms are called "coherence conditions."

In examples 2.6.1 and 2.6.2 the existence of c is obvious; for example, in (2.6.2) we take

$$c = \begin{bmatrix} O & I \\ I & O \end{bmatrix} : R^n \oplus R^m \longrightarrow R^m \oplus R^n .$$

It is also possible to make a similar weakening of the notion of a "strict monoidal category" in which the product $\square$ is associative, and has a unit, up to coherent isomorphisms only. One then speaks of a "monoidal category"; see [84] pp. 158-159.

Some authors say "permutative category" instead of "symmetric strict monoidal category" [93].

Next I must explain that there is a little gadget for taking the "classifying space" of a category. Suppose given a (small) category C; an appropriate example to consider is that in which the category C has one object X, and the morphisms from X to X form a group; but it could be any small category. Regard the object $\{0,1,2,\cdots,n\}$ in Δ as itself a category, namely the category

$$0 \leftarrow 1 \leftarrow 2 \leftarrow \cdots \leftarrow n$$

in which there is one map from i to j if $i \geq j$. (Some authors regard it as the category

$$0 \rightarrow 1 \rightarrow 2 \rightarrow \cdots \rightarrow n ,$$

and it ought not to make any difference in the end, but somehow if I do it that way I cannot get all my formulae consistent; I expect it comes from writing my maps on the left of their arguments.) Take then

$$K_n = \mathrm{Hom}(0 \leftarrow 1 \leftarrow 2 \leftarrow \cdots \leftarrow n, \ C) ,$$

the set of functors from $0 \leftarrow 1 \leftarrow 2 \leftarrow \cdots \leftarrow n$ to C. This is like taking "1-cocycles on σ^n with values in C" (compare §2.5). A morphism in Δ is just the same as a functor from $0 \leftarrow 1 \leftarrow 2 \leftarrow \cdots \leftarrow n$ to $0 \leftarrow 1 \leftarrow 2 \leftarrow \cdots \leftarrow m$, so it induces a function $K_n \leftarrow K_m$, and we get a simplicial set K. This simplicial set is called the "nerve" of the category; let us write it Nerve(C).

The definition proposed by Segal [124] for the "classifying space of a category" is now

$$BC = |\mathrm{Nerve}(C)| ;$$

we take Nerve(C) and take its geometric realization (see §2.4).

If the category C reduces to one object provided with a discrete group π of automorphisms, then Nerve(C) reduces to the simplicial set $K(\pi,1)$ introduced by Eilenberg-MacLane and BC is the best model for $B\pi$, the classifying space for π, which should be an Eilenberg-MacLane space of type $(\pi,1)$.

In examples 2.6.1 and 2.6.2, the category C reduces to a sequence of discrete groups G_n, and BC reduces to

$$\coprod_{n \geq 0} BG_n .$$

If we have a bifunctor $\square$ which makes C into a strict monoidal category, then by exploiting it appropriately we can make BC into a strict monoid.

The appropriate general result is now as follows.

PRETHEOREM 2.6.3. *If* C *is a symmetric monoidal category, then* BC *is an* E_∞*-space.*

See [93] p. 78.

The reason one might wish to do without the word "strict" here is that one might treat some examples more conveniently that way. Suppose we wished to make a version of example 2.6.2 in which the objects of C are the finitely generated projective R-modules and the morphisms are their isomorphisms; then we might prefer not to have to make it into a *strict* monoidal category. However, since we can always arrange "strictness" by an appropriate construction on categories [72] it seems unreasonable to fuss.

I remark that examples 2.6.1 and 2.6.2 serve to show that some of the generality in §2.3 is not superfluous. In these examples BC is an E_∞-space, and so we can use it to construct a spectrum $\mathbf{Y} = B^\infty(BC)$. However, $\Omega^\infty \mathbf{Y}$ is not equivalent to BC, because $\pi_0(BC)$ is not a group; in (2.6.1) and (2.6.2) $\pi_0(BC)$ is $\{0,1,2,3,\cdots\}$, the monoid of non-negative integers under addition. So the relation between BC and $\Omega^\infty \mathbf{Y}$ will be as discussed in §3.2.

Promising further discussion in §3.2, I simply state that in example 2.6.1 we have

$$BC = \coprod_{n \geq 0} B\Sigma_n ,$$

$$\mathbf{Y} = \Sigma^\infty S^0 ,$$

$$\Omega^\infty \mathbf{Y} = \Omega^\infty \Sigma^\infty S^0 = \lim_{n\to\infty} \Omega^n S^n ;$$

while example 2.6.2 leads to algebraic K-theory.

It remains to say that I have slightly oversimplified the theory as it exists. For example, in (2.6.2) I used a discrete ring R; one might also wish to use a topological ring R, such as the real or complex numbers, and so bring topological K-theory into the scope of this theory. For this purpose, of course, the category C will have to be a topological category. We may as well go the whole hog and assume that both $\mathrm{Ob}(C)$, the set of objects of C, and $\mathrm{Mor}(C)$, the set of morphisms of C, have topologies, so that the four structure maps

$$\begin{aligned} \text{source: } & \mathrm{Mor}(C) \longrightarrow \mathrm{Ob}(C) \\ \text{target: } & \mathrm{Mor}(C) \longrightarrow \mathrm{Ob}(C) \\ \text{identity: } & \mathrm{Ob}(C) \longrightarrow \mathrm{Mor}(C) \\ \text{composition: } & \mathrm{Mor}(C) \times_{\mathrm{Ob}(C)} \mathrm{Mor}(C) \longrightarrow \mathrm{Mor}(C) \end{aligned}$$

are continuous. This makes no great difference to the construction of BC, since in §2.4 we defined the geometric realization of simplicial spaces as well as simplicial sets. It should not affect the truth of Pretheorem 2.6.3.

§2.7. *Ring-theories*

In this section I will offer a few inadequate remarks about multiplicative structures.

In algebraic topology one generally feels that the more algebraic structure one can get, the better. In particular one has the notion of a "generalized cohomology theory with products"; if k^* is such a theory one is given cup-products

$$k^p(X) \otimes k^q(X) \longrightarrow k^{p+q}(X) ;$$

or if one prefers external products, one may be given

$$k^p(X) \otimes k^q(Y) \longrightarrow k^{p+q}(X \times Y)$$

or

$$\widetilde{k}^p(X) \otimes \widetilde{k}^q(Y) \longrightarrow \widetilde{k}^{p+q}(X \wedge Y) .$$

Those who do stable homotopy theory have developed notions for spectra which correspond to these notions for cohomology theories. It is necessary to define the smash-product $E \wedge F$ of two spectra; once this is done, one defines a "ring-spectrum" to be a spectrum E which comes equipped with a product map

$$\mu : E \wedge E \to E$$

satisfying suitable axioms. Then such familiar spectra as MU are ring-spectra; and from a ring-spectrum we get a cohomology theory with products. For these points, see [9]. However, I must point out that the "smash-product" of spectra constructed there, while adequate for the purposes described above, is not licensed as suitable for more ambitious purposes.

Other examples certainly show that it is reasonable to consider a space X equipped with two separate and distinct structure maps $\alpha, \mu : X \times X \to X$, which we think of as "addition" and "multiplication."

For example, take $X = \Omega^n S^n$; it has an addition map $\alpha : X \times X \to X$ obtained by considering X as $\Omega(\Omega^{n-1} S^n)$ and using "addition of loops" as one does in ΩY for any Y; it also has a multiplication map $\mu : X \times X \to X$ obtained by considering X as the set of based maps from S^n to S^n and using composition of maps. More accurately, one would wish to take this example and pass to the limit as $n \to \infty$.

A space X with two such structure maps α, μ may be called a "ring-space." More precisely, X will be a "ring-object in the homotopy category" if α and μ make $[W,X]$ into a functor from spaces W to rings. This of course demands assumptions (up to homotopy) and α and μ both separately and together; for example, the distributive law must hold up to homotopy.

In particular, let E be a ring-spectrum; then the 0th cohomology group $E^0(W)$ is a functor from spaces to rings, and therefore the space

$\Omega^\infty E$ is a ring-object in the homotopy category. For example, if our cohomology theory is K-theory, we see that the space $X = Z \times BU$ is a ring-space; the addition map $\alpha : X \times X \to X$ corresponds to the Whitney sum of vector-bundles, and the multiplication map $\mu : X \times X \to X$ corresponds to the tensor-product of vector-bundles.

However we can go further, and make demands on α and μ not merely "up to homotopy," but "up to an infinity of higher homotopies," in a sense modelled on that sketched in §2.2, §2.3. We thus reach the notion of an "E_∞ ring space."

For example, suppose given a category C with two compatible monoidal structures, along the lines of §2.6; thus, we might envisage taking C to be the category of finite sets with "disjoint union" and "Cartesian product" as the two operations. One would hope that the space BC constructed in §2.6 would provide an E_∞ ring space.

The notion of E_∞ ring space should be the starting-point of an interesting and very rich theory. The reader might hope for some analogue of Pretheorem 2.3.2, which showed that the study of E_∞ H-spaces X was essentially the same as the study of spectra. However, the spectral counterpart of an E_∞ ring space should be a spectrum with infinitely much extra structure. There is a natural way to begin: one would naturally write down the form of words "a ring-spectrum E in which the product map $\mu : E \wedge E \to E$ is commutative and associative up to an infinity of higher homotopies," and one would continue by trying to make sense of this form of words. For this purpose it would be highly desirable to define the smash product of spectra in a way more nearly ideal, so that it is as nearly associative as the cartesian product of spaces is. To do so one has to re-examine the foundations of the theory of spectra. This has been done, and the outcome is the notion of "coordinate-free spectra"; see [99] Chapter 2. In an ordinary spectrum, we are given one space E_n for each integer n. In a coordinate-free spectrum, we are given one space $E(V)$ for each n-dimensional subspace V of R^∞, but for suitable isomorphisms $V \to V'$ we are given homeomorphisms $E(V) \to E(V')$.

This has the effect of cutting the data back to the same essential size; but it also affords an automatic machine for keeping control of the nuisance which would otherwise be caused by permuting suspension coordinates. I am fully persuaded that coordinate-free spectra are a Good Thing; they make smash products much nicer and allow one to keep smash products under the requisite control; but I do not want to spend the space to go into that here. Once one has the technical details of coordinate-free spectra, one may use them to write down the definition of an "E_∞ ring spectrum."

It is claimed, and it is plausible, that E_∞ ring spectra are useful when one wants to study questions which involve a very precise hold over the product structures, such as questions on the orientability of spherical fibrations over generalized cohomology theories.

There is also the slightly weaker notion of an "H_∞ ring spectrum." This is an analogue for spectra of the notion of a space provided with structure maps

$$(E\Sigma_n)\times_{\Sigma_n}(X \wedge X \wedge \cdots \wedge X) \longrightarrow \Sigma^? X ,$$

subjected to axioms saying that certain diagrams commute up to homotopy. Such a notion may be attractive to homotopy-theorists; it can be used to provide an appropriate context in which to prove, for example, Nishida's nilpotency theorem [113]. For H_∞ ring spectra, see [98].

I will close this section by simply affirming a few points. I believe that theories of the sort indicated can be developed, have been developed, and are worth-while; it must be right to carry a multiplicative structure around with us whenever we can. However I am not yet ready to report more fully; and I refer the reader to [99, 98].

CHAPTER 3

LOCALIZATION AND "GROUP COMPLETION"

§3.1. *Localization*

The theme of this chapter is that one space can be like another without actually being equivalent to it. I want to describe two constructions; both of them take a space X and construct another space which is related to X without being equivalent to X. I begin with localization, which is of very general use in topology.

In homotopy-theory, the idea that we can consider 2-primary problems independently of 3-primary problems goes back to Serre [130]. In algebra, there is a functorial construction for retaining 2-primary problems and ignoring 3-primary, 5-primary and other problems. This is the device of localization. For example, if your problem concerns Z-modules, you can define $Z_{(2)}$, the integers localized at 2, to be the set of fractions $^a/_b \in Q$ such that b is odd; then if M is a Z-module you define its localization at 2 by $M_{(2)} = M \otimes_Z Z_{(2)}$. This has the effect of retaining the 2-primary problems and suppressing 3-primary, 5-primary and other problems. Or if you just want to suppress the 2-primary problems and retain all the others you can form $M \otimes_Z Z[½]$.

It seems that Sullivan [149, 150] was the first to see with real clarity that in homotopy-theory there is a functorial construction analogous to localization in algebra. Of course Sullivan's insights had a wide influence before they were published; there were also mathematicians who worked independently of Sullivan and published quite early.

We may approach the subject by asking how one would introduce coefficients into a generalized homology theory. So, let A be an abelian group. Then our way is quite clear. We first construct a Moore space Y

– that is, a space with one non-vanishing homology group, that group being A. For example, choose a resolution of A by free Z-modules, say

$$0 \longrightarrow R \xrightarrow{d} F \longrightarrow A \longrightarrow 0\,;$$

then by taking suitable wedges of spheres, we can arrange

$$H_n\Big(\bigvee_{\beta \in B} S^n\Big) \cong R$$

$$H_n\Big(\bigvee_{\gamma \in \Gamma} S^n\Big) \cong F\,;$$

and we can arrange a map

$$f\colon \bigvee_{\beta \in B} S^n \longrightarrow \bigvee_{\gamma \in \Gamma} S^n$$

such that the following diagram is commutative.

$$\begin{array}{ccc} H_n\Big(\bigvee_{\beta \in B} S^n\Big) & \xrightarrow{f_*} & H_n\Big(\bigvee_{\gamma \in \Gamma} S^n\Big) \\ \Big\updownarrow \cong & & \Big\updownarrow \cong \\ R & \xrightarrow{d} & F \end{array}$$

We have only to take

$$Y = \Big(\bigvee_{\gamma \in \Gamma} S^n\Big) \cup_f C\Big(\bigvee_{\beta \in B} S^n\Big).$$

Suppose now that we are given a generalized homology theory k_*. Then we can define the corresponding theory with coefficients by

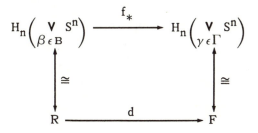

$$\widetilde{k}_m(X;\, A) = \widetilde{k}_{n+m}(X \wedge Y)\,.$$

It follows that we have the universal coefficient theorem in the form of a short exact sequence

$$0 \longrightarrow \widetilde{k}_m(X) \otimes_Z A \longrightarrow \widetilde{k}_m(X;\, A) \longrightarrow \mathrm{Tor}_Z(\widetilde{k}_{m-1}(X), A) \longrightarrow 0\,.$$

From this it is perfectly clear how we should introduce coefficients into a spectrum. That is, if we have a spectrum E, we should define a new spectrum F by

$$F_m = E_{m-n} \wedge Y\,.$$

If we specialize to the case in which A is torsion-free, then we have isomorphisms

$$\pi_r(F) \cong \pi_r(E) \otimes A$$

$$H_r(F) \cong H_r(E) \otimes A\,.$$

We have thus defined a functor on spectra, which on their homotopy groups π_r and on their homology groups H_r has the effect "$\quad \otimes A$."

The question is now, whether we can find a construction on spaces which on infinite-loop-spaces will agree with this construction on spectra. I say "spaces," but it will be sufficient to deal with simply-connected CW-complexes; the theory can be made to tolerate a small amount of nuisance from the fundamental group, but not too much, and I will omit discussion of this point.

The case in which A is a subring of the rationals goes in a most satisfactory way. In this case, we say that a Z-module M is "A-local" if it can be given the structure of an A-module. (If so, then that structure is unique; it also happens that any Z-map between A-modules is an A-map.) Let X be a 1-connected CW-complex; we say that X is "A-local" if (i) its homotopy groups $\pi_r(X)$ are A-local, or equivalently (ii) its homology groups $H_r(X)$ are A-local.

THEOREM 3.1.1 (Sullivan [149] p. 18). *Let* X *be a* 1-*connected* CW-*complex and* A *a subring of the rationals. Then there is a* 1-*connected* CW-*complex* Y *and a map* $i: X \to Y$ *with the following properties.*

(i) Y *is* A-*local and the map* $i: X \to Y$ *is universal among maps into* A-*local spaces.*

(ii) i *localizes homotopy, in the sense that*

$$i_*: \pi_r(X) \longrightarrow \pi_r(Y)$$

induces

$$\pi_r(X) \otimes A \xrightarrow{\cong} \pi_r(Y) .$$

(iii) i *localizes homology, in the sense that*

$$i_*: H_r(X) \longrightarrow H_r(Y)$$

induces

$$H_r(X) \otimes A \xrightarrow{\cong} H_r(Y) .$$

We shall write XA for such a space Y, and regard it as "X localized so as to introduce coefficients in A." For example, $XZ[\frac{1}{2}]$ means X localized so as to suppress 2-primary problems, and so on.

Let me sketch the two methods which Sullivan gives for constructing Y. First we note that if X happens to be a sphere S^n, then the outcome is forced by clause (iii); we must take Y to be a Moore space of type (A,n). But any CW-complex X can be constructed by successively attaching cells CS^n; so we copy this process and construct Y by successively attaching cones on Moore spaces. Of course in this process we need suitable maps by which to attach our cones, but that's what you have inductive hypotheses and lemmas for.

The second method is dual. We note that if X happens to be an Eilenberg-MacLane space of type (π,n), then the outcome is forced by clause (ii); we must take Y to be an Eilenberg-MacLane space of type $(\pi \otimes A, n)$. But any space X has a "Postnikov decomposition," that is, X can be decomposed as an iterated fibering in which the building-blocks are Eilenberg-MacLane spaces; we then construct Y by induction up the Postnikov decomposition of X, localizing all the spaces in that decomposition as we go. The last sentence of the previous paragraph dualizes too.

For later use, it will be convenient to write "$EM(\pi,n)$" for an Eilenberg-MacLane space of type (π,n).

Next I must say something about the place which localization now has in topology. We need it partly as a matter of language, to state results which without localization either could not be stated at all, or else could only be stated in a less convenient form. (It goes without saying that results so stated can have consequences which can be stated without localization.) As an example of the first type, one might quote a statement from Sullivan [149] p. 24; we have

$$(F/Top)\, Z[\tfrac{1}{2}] \cong BO\, Z[\tfrac{1}{2}]$$

$$(F/Top)\, Z_{(2)} \cong \prod_n EM(\pi_n, n),$$

where π_n is as follows:

$$\begin{array}{cccccc} n \equiv 0 & 1 & 2 & 3 & 4 & \text{mod } 4 \\ \pi_n = Z_{(2)} & 0 & Z_2 & 0 & Z_{(2)}\,. & \end{array}$$

Since the 2-primary behavior of F/Top is different from its behavior at odd primes, it is hardly credible that one can describe F/Top without separating the primes. The applications are to the theory of manifolds.

Again, Theorem 6.2.1 below (the theorem of Adams and Priddy) is stated using localization, and would not be true without localization.

As an example of the second type, Theorem 5.1.1 below (the "Adams conjecture") was originally stated without localization, but the statement which does use localization is more convenient and illuminating. The original application was to the study of the J-homomorphism in homotopy theory.

Once we have a genuine need for some mathematical idea as a matter of language, that idea has arrived; it is hardly necessary to discuss its status as a useful technical tool. (That sentence is not intended to exclude the possibility that some authors may try to introduce language we can do without.) Nevertheless, I would like to illustrate the use of localization in constructing certain examples.

Localization can be used to give examples of "finite H-spaces" which are not classical. For example, one can construct a finite complex X which is an H-space, and which to 2-primary eyes looks like Sp(2), but to 3-primary eyes looks like $S^3 \times S^7$. Then X is not Sp(2) (look at it with 3-primary eyes) and equally it is not $S^3 \times S^7$ (look at it with 2-primary eyes). This is perhaps the crudest example of a technique known as "Zabrodsky mixing" [156]. It is somewhat reminiscent of the descriptions of monsters in medieval bestiaries; you put the head of a lion on the body of a horse. A more subtle technique is to take parts from the same animal and put them back together a different way round; this is as if you take your lion, cut his head off and then stick the same head back on, but facing to the rear. This is what happens in the Hilton-Roitberg example [67, 68]; the part of the lion is played by Sp(2).

As a further example, we can assign the lion's role to HP^∞, the infinite-dimensional projective space over the quaternions. I state that there is a 3-connected CW-complex X with the following three properties. (i) $\Omega X \simeq S^3$. (ii) For each prime p, $XZ_{(p)} \simeq HP^\infty Z_{(p)}$. These first two properties say that X is like HP^∞; spaces with these properties have been studied by Rector [121]. But the third property goes in the opposite direction. Recall (e.g. from §1.8) that any map $f: X \to BU$ has a total Chern class. (iii) If the total Chern class of f is finite, then it is 1. This says that there is a severe shortage of finite-dimensional vector bundles over X, and in this respect X is not like HP^∞.

Of course, when one comes to use localization, one needs results about it which I have not yet mentioned. Theorem 3.1.1 is perfectly satisfactory as long as we want to pass from spaces to their localizations; but sometimes we want to go in the reverse direction, by taking data about the localized spaces and assembling it to obtain global conclusions, that is, conclusions about the original, unlocalized spaces. At the risk of seeming superficial, I select two results which it is perhaps better for the reader to have seen rather than not.

The first result is useful when we wish to assemble information from two sets of primes. Let A,B be subrings of the rationals; then we can form the following diagram.

$$\begin{array}{ccc} A \cap B & \longrightarrow & A \\ \downarrow & & \downarrow \\ B & \longrightarrow & A \otimes B \end{array}$$

PROPOSITION 3.1.2. *For each 1-connected CW-complex X the diagram*

$$\begin{array}{ccc} X(A \cap B) & \longrightarrow & XA \\ \downarrow & & \downarrow \\ XB & \longrightarrow & X(A \otimes B) \end{array}$$

is both a fibre square and a cofibre square. Moreover, if W is a finite 1-connected CW-complex, then a map

$$W \xrightarrow{f} X(A \cap B)$$

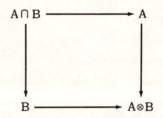

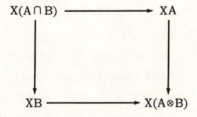

is determined up to homotopy by the two composites

$$W \longrightarrow X(A\cap B) \begin{array}{l} \nearrow XA \\ \searrow XB \end{array} ;$$

that is, if we only map in finite complexes W, *then we have the strict pull-back property.*

The second result is useful when we wish to assemble information which comes from each prime p separately.

PROPOSITION 3.1.3. (a) *Let* W *and* X *be* 1-*connected* CW-*complexes, with* W *finite. Then a map*

$$W \xrightarrow{f} X$$

is determined up to homotopy by the composites

$$W \xrightarrow{f} X \longrightarrow XZ_{(p)}$$

(*where* p *runs over all primes*).

(b) *Assume further that the homotopy groups of* X *are finitely-generated, at least in dimensions* $\leq \dim W$. *Suppose given maps*

$$W \xrightarrow{f_p} XZ_{(p)}$$

(*one for each prime* p) *such that the composites*

$$W \xrightarrow{f_p} XZ_{(p)} \longrightarrow XQ$$

all lie in a common homotopy class independent of p. *Then there exists a map*

$$W \xrightarrow{f} X$$

such that the composite

$$W \xrightarrow{f} X \longrightarrow XZ_{(p)}$$

is homotopic to f_p *for each* p.

The restrictive conditions in these propositions cannot be avoided.

For standard references and further reading about localization, I suggest [45, 46, 64, 65, 66, 97, 109].

§3.2. *The "plus construction" and "group completion"*

The next situation we want to consider is the following. We have two spaces X, Y and a map $f: X \to Y$ which induces isomorphisms of homology groups, but yet the fundamental groups $\pi_1(X)$ and $\pi_1(Y)$ are wildly different, and therefore the higher homotopy groups have to be wildly different too to get the homology the same. Alternatively, it may be that $f_*: H_*(X; A) \to H_*(Y; A)$ is an isomorphism only for suitable coefficients A.

The first example of this situation of which I became aware was provided by Quillen [118]. In his context the space Y was the space BU. To construct the space X, we begin by taking a prime p. Let F_p be the finite field with p elements, and let $\bar{F}_p$ be the algebraic closure of F_p. Let

$$GL(\infty, \bar{F}_p) = \bigcup_n GL(n, \bar{F}_p)$$

be the infinite-dimensional linear group over $\bar{F}_p$; we consider it as a discrete group because there is no other sensible way to put a topology on it. We take X to be the classifying space $BGL(\infty, \bar{F}_p)$. This then is an Eilenberg-MacLane space, whose fundamental group is this great big non-abelian group $GL(\infty, \bar{F}_p)$, and whose higher homotopy groups are zero.

Nevertheless, Quillen constructed a map

$$f\colon BGL(\infty, \bar{F}_p) \longrightarrow BU$$

and showed that the induced homomorphism

$$f^*\colon H^*(BGL(\infty, \bar{F}_p); A) \longleftarrow H^*(BU; A)$$

is an isomorphism whenever A is a finite group with no p-torsion. This meant that for the rest of Quillen's argument, $BGL(\infty, \bar{F}_p)$ was an entirely acceptable substitute for BU, because the argument was by obstruction-theory. I shall explain the context and purpose of Quillen's argument later (see Chapter 5).

Quillen also treated the case $Y = BO$, $X = BO(\infty, \bar{F}_p)$ (p odd).

Quillen constructed his map by representation-theory, using theorems going back to R. Brauer and J. A. Green. Given a good map, one would almost be willing to prove that it induces an isomorphism of cohomology by direct calculation of both cohomology groups; and what Quillen did was not far off that, although it was more elegantly expressed.

If one wants to seek the real reason why this result is true (although the question may be thought theological), one should probably look in algebraic geometry and étale homotopy theory. However this surely does not apply to the second example of which I became aware.

For this example, let

$$\Sigma_\infty = \bigcup_n \Sigma_n$$

be the infinite symmetric group; we take $X = B\Sigma_\infty$ and

$$Y = \Omega^\infty \Sigma^\infty S^0 = \lim_{n \to \infty} \Omega^n S^n .$$

The result for this case was first published, I think, by Priddy [114].

However almost everyone who has worked in this area seems to have his own favorite treatment of the situation. An elementary proof is simple enough: one can write down the map

$$f: B\Sigma_\infty \longrightarrow \Omega^\infty\Sigma^\infty S^0$$

and check by force if necessary that

$$f_*: H_*(B\Sigma_\infty; A) \longrightarrow H_*(\Omega^\infty\Sigma^\infty S^0; A)$$

is an isomorphism for all coefficients A.

If one wants to seek the real reason why this result is true (modulo the same disclaimer as before) is is perhaps this: when you construct a model for $\Omega^\infty\Sigma^\infty S^0$, by the method of J. P. May or any other available method, you use the spaces $B\Sigma_n$ and nothing else. However, this remark also begs the question at issue, namely, in what sense the model approximates to $\Omega^\infty\Sigma^\infty S^0$.

By now we have theorems saying that this situation is common; Kan and Thurston [76] show that given almost any space Y, you can approximate it homologically by an Eilenberg-MacLane space $EM(\pi,1)$ for some weird and artificial group π. However, we should perhaps be more concerned with cases where this situation arises in nature.

I come now to the first suggestion for describing this situation. I believe that it was due to Quillen [117] and that it was directly inspired by the need to find a topological interpretation for Milnor's functor $K_2(R)$. Let X be a space, and let π be a normal subgroup of the fundamental group $\pi_1(X)$ which is perfect in the sense that $[\pi,\pi] = \pi$. For a first example, we can take X to be $B\Sigma_\infty$ and π to be the alternating subgroup of Σ_∞. For a second example, we can take X to be $BGL(\infty, R)$ and π to be the subgroup of $GL(\infty, R)$ generated by the elementary matrices. Then there is a way to construct a space X^+ and a map $i: X \to X^+$ to satisfy the following conditions.

(i) $i_*: \pi_1(X) \to \pi_1(X^+)$ is an epimorphism with kernel π.

(ii) Let $\pi_1(X^+)$ act on any abelian group A, so that we get a local coefficient system over X^+ and one over X. Then

$$i^*: H^*(X; A) \longleftarrow H^*(X^+; A)$$

is iso.

This construction is known as Quillen's "plus construction"; it has the effect of cutting a great big fundamental group down to size without affecting the cohomology groups. It is usual not to give the construction in detail; in the original source Quillen was delivering three lectures in the time meant for one, and subsequent authors have not liked to take up space with something already known. (However, one must commend Wagoner [154] for being kind to his readers.) I will take the space to explain.

The idea is to add 2-cells to kill π and 3-cells which neutralize the 2-cells so far as cohomology goes. In more detail, we can assume that X is a CW-complex. Let $p: \widetilde{X} \to X$ be the regular covering corresponding to the normal subgroup $\pi \subset \pi_1(X)$. Take any element $x \in \pi$; since π is perfect we can write it in the form

$$x = \prod_{i=1}^{m} [y_i, z_i],$$

where $y_i, z_i \in \pi$. Let X^1, $\widetilde{X}^1$ be the 1-skeletons of X, $\widetilde{X}$; since the map

$$\pi_1(\widetilde{X}^1) \longrightarrow \pi_1(\widetilde{X}) = \pi$$

is epi, we can choose elements $\bar{y}_i$, $\bar{z}_i$ in $\pi_1(\widetilde{X}^1)$ which map to y_i, z_i in π; we can form a new complex Y by attaching to X a 2-cell, using an attaching map in the class

$$p_* \prod_{i=1}^{m} [\bar{y}_i, \bar{z}_i].$$

The covering $\widetilde{X}$ of X extends to a covering $\widetilde{Y}$ of Y; here $\widetilde{Y}$ is obtained from Y by attaching a set of 2-cells, one of which has an attaching map in the class

$$\prod_{i=1}^{m} [\overline{y}_i, \overline{z}_i],$$

while the others are the translates of the first under the group of covering translations $G = \pi_1(X)/\pi$.

We proceed in this way for a set of elements $x \in \pi$ sufficient to generate π as a normal subgroup; let us use the same letters $Y, \widetilde{Y}$ for the result. Then $\widetilde{Y}$ is simply-connected. By the Hurewicz theorem, we have

$$\pi_2(\widetilde{Y}) \cong H_2(\widetilde{Y}).$$

Since the attaching maps we used in $\widetilde{Y}$ were nullhomologous in $\widetilde{Y}^1$, we have

$$H_2(\widetilde{Y}) \cong H_2(\widetilde{X}) \oplus F$$

where F is a free module over $Z[G]$ (on generators corresponding to the cells we added to X). Let $f_\alpha \in \pi_2(\widetilde{Y}^2)$ be elements which map to a $Z[G]$-base for F. We construct X^+ by attaching 3-cells to Y, using attaching maps in the classes $p_* f_\alpha$. We may describe the covering $\widetilde{X}^+$ as we described $\widetilde{Y}$; for each α we have one 3-cell with an attaching map in the class f_α, and also the translates of that cell under G. If we use cellular chains, then the relative chain complex $C_*(\widetilde{X}^+, \widetilde{X})$ becomes

$$\cdots 0 \longrightarrow F \xrightarrow{\cong} F \longrightarrow 0 \longrightarrow 0 .$$

Thus the injection $C_*(\widetilde{X}) \to C_*(\widetilde{X}^+)$ is a chain equivalence over $Z[G]$. This completes the proof of properties (i), (ii) above.

We may now give Quillen's first construction of the higher groups of algebraic K-theory, as follows. Take the space $BGL(\infty, R)$, and perform the plus construction, taking π to be the subgroup of $GL(\infty, R)$ generated by the elementary matrices. Define

$$K_i(R) = \pi_i((BGL(\infty, R))^+) .$$

Of course, for any such application it is important to know that the conditions (i), (ii) above do characterize X^+ up to canonical homotopy equivalence. This is proved by an easy argument, using obstruction-theory. In this application of condition (ii) we take the coefficient system A to be the homotopy groups of one candidate for X^+ —let us write $\pi_n(X^+)$. Now in all cases of interest X^+ is an H-space, so that the action of $\pi_1(X^+)$ on $\pi_n(X^+)$ is trivial. We therefore discern the possibility of avoiding further fuss and nuisance about local coefficients.

Unfortunately, our data so far certainly do not imply that X^+ is an H-space, or even that the action of $\pi_1(X^+)$ on $\pi_n(X^+)$ is trivial. But this is something we can check directly in our examples. Consider for example the case $X = B\Sigma_\infty$. We can construct a homomorphism $\mu: \Sigma_\infty \times \Sigma_\infty \to \Sigma_\infty$ by letting the first copy of Σ_∞ permute the odd integers while the second copy of Σ_∞ permutes the even integers. The classifying-space construction preserves products, so we get a map

$$B\mu\colon B\Sigma_\infty \times B\Sigma_\infty \longrightarrow B\Sigma_\infty .$$

Now this map does not make $X = B\Sigma_\infty$ into an H-space, because the base point does not act as a unit. In fact $B\Sigma_\infty$ cannot be an H-space, because its fundamental group Σ_∞ is not abelian. However, the plus construction is functorial and preserves products. Therefore we get a map

$$(B\mu)^+\colon (B\Sigma_\infty)^+ \times (B\Sigma_\infty)^+ \longrightarrow (B\Sigma_\infty)^+ .$$

We now check that the restrictions of $(B\mu)^+$ to $(B\Sigma_\infty)^+ \times pt$ and $pt \times (B\Sigma_\infty)^+$ are homotopy equivalences. Therefore, even if $(B\mu)^+$ is not an

H-space structure map, we can replace it by a product map

$$\mu' : (B\Sigma_\infty)^+ \times (B\Sigma_\infty)^+ \longrightarrow (B\Sigma_\infty)^+$$

which has the base-point as a strict unit. Therefore the action of $\pi_1((B\Sigma_\infty)^+)$ on $\pi_n((B\Sigma_\infty)^+)$ is trivial.

The case $X = BGL(\infty, R)$ is similar; we can construct a homomorphism

$$\mu: GL(\infty,R) \times GL(\infty,R) \longrightarrow GL(\infty,R)$$

by letting the first matrix act on the odd-numbered base vectors and the second matrix act on the even-numbered base vectors; then we argue as above. Indeed we could no doubt axiomatize the situation; the obvious axiomatization would be for $X = BG_\infty$, where $G_\infty = \bigcup_n G_n$ and the groups G_n form a permutative category.

In this sort of case, then, we can characterize X^+ by the following statements.

(i) $i_* : \pi_1(X) \to \pi_1(X^+)$ is an epimorphism with kernel π.

(iii) The action of $\pi_1(X^+)$ on $\pi_n(X^+)$ is trivial for each n.

(iv) $i_* : H_*(X) \to H_*(X^+)$ is iso (with constant coefficients Z).

I come now to my second topic. In §2.3 I promised the reader something; more precisely, a description of the relation between M and ΩBM when M is a topological monoid which does not need to be connected. Of course we have a map $M \to \Omega BM$, and we would like to describe the homological behavior of this map. Typically it does not induce an isomorphism of the homology of any single component of M, but you do obtain an isomorphism by passing to a limit over the components of M. After giving some general definitions we can look at some examples.

If X is an H-space, then the product in X defines a product in $\pi_0(X)$. We say that X is "grouplike" if this product makes $\pi_0(X)$ a group. For example, any loop-space is grouplike, because $\pi_0(\Omega Y) \cong \pi_1(Y)$. We are not willing to assume that our given monoid M is grouplike. In fact, our examples are

(a) $$M = \coprod_{n \geq 0} B\Sigma_n ,$$

(b) $$M = \coprod_{n \geq 0} BGL(n, R) .$$

In these examples $\pi_0(M) = \{0,1,2,\cdots\}$ under addition. It seems safest not to restrict ourselves to this particular monoid, because one might have other applications; for example, in the context of algebraic K-theory one might wish to start from the category C of finitely-generated projective R-modules and their isomorphisms, and in that case $\pi_0(M)$ would turn out to be the "Grothendieck monoid" of isomorphism classes of finitely-generated projective R-modules (under addition).

We will write α for a typical element of $\pi_0(M)$ and M_α for the corresponding path-component of M, so that in examples (a) and (b) above, M_n is $B\Sigma_n$ or $BGL(n, R)$ for $n = 0,1,2,\cdots$.

We now form BM, the classifying space of the monoid M, and ΩBM, the loop-space on BM. We have a map

$$i : M \to \Omega BM$$

and this is a map of H-spaces. It is easy to see what it does to π_0; the group $\pi_0(\Omega BM)$ is the universal group associated to the monoid $\pi_0(M)$, and the map

$$i_*: \pi_0(M) \longrightarrow \pi_0(\Omega BM)$$

has the universal property.

In examples (a) and (b) above we have $\pi_0(M) = \{0,1,2,\cdots\}$ and $\pi_0(\Omega BM) = Z$. However, if we started from the category C of finitely-generated projective R-modules and their isomorphisms we would get

$$\pi_0(\Omega BM) = K_0(R) .$$

Since ΩBM is a grouplike H-space its path-components are all homotopy-equivalent; we have

$$\Omega BM \cong \pi_0(\Omega BM) \times (\Omega BM)_0 .$$

However the path-components of M need not be homotopy-equivalent; in examples (a), (b) above we have

$$M_n = B\Sigma_n, \quad M_n = BGL(n,R) .$$

It is clear that in these examples we are interested in the limit of M_n as n tends to infinity, that is in $B\Sigma_\infty$ or $BGL(\infty, R)$.

THEOREM 3.2.1 ("Group completion theorem," [21, 27, 93, 94, 101, 120, 127]). *Under suitable assumptions,*

$$\varinjlim_{\alpha \in \pi_0(M)} H_*(M_\alpha) \longrightarrow H_*((\Omega BM)_0)$$

is iso; equivalently,

$$Z[\pi_0(\Omega BM)] \otimes_{Z[\pi_0(M)]} H_*(M) \longrightarrow H_*(\Omega BM)$$

is iso.

Before explaining the assumptions and the conclusion, I will comment on the use of the theorem.

COROLLARY 3.2.2. *In the two examples,*

$$M_n = B\Sigma_n \quad \textit{and} \quad M_n = BGL(n,R) ,$$

we have

$$(\Omega BM)_0 = (B\Sigma_\infty)^+, \qquad (\Omega BM)_0 = (BGL(\infty,R))^+ .$$

More broadly, we can say that the theorem proves the agreement of two approaches to the higher K groups $K_i(R)$. The first is Quillen's first approach in terms of the plus construction, and the definition is $K_i(R) = \pi_i((BGL(\infty,R))^+)$. The second approach takes the category of finitely-generated projective R-modules and feeds it into a suitable machine; the outcome is an Ω-spectrum with 0th term ΩBM; the groups $k_i(R)$ appear as the homotopy groups of the spectrum. The plus construction is more elementary, but what it gives is a bare homotopy type; the 0th term of an Ω-spectrum comes with a much richer structure.

Sketch proof of Corollary 3.2.2. Let M_∞ be $\lim_{n\to\infty} M_n$, that is $B\Sigma_\infty$ or $BGL(\infty, R)$ according to the case; it is easy to construct a map from it into $(\Omega BM)_0$, and this map induces an isomorphism of homology by the theorem. Since $(\Omega BM)_0$ is an H-space we have

$$\begin{aligned} \pi_1(\Omega BM)_0 &= H_1(\Omega BM)_0 \\ &\cong H_1(M_\infty) \\ &\cong \frac{\pi_1(M_\infty)}{[\pi_1(M_\infty), \pi_1(M_\infty)]} \quad ; \end{aligned}$$

this checks the condition on the fundamental group. It is now clear that $(\Omega BM)_0$ satisfies our second characterization (i) (iii) (iv) of X^+, where $X = M_\infty$.

From one point of view it is quite tempting to try and strengthen the group completion theorem so as to assert a conclusion about cohomology with local coefficients; this would allow one to check that $(\Omega BM)_0$ satisfies the first characterization (i) (ii) of X^+; but we can do without that and I know no great advantage of such a strengthening.

We must now explain the statement and the assumptions of the theorem. The second form of the statement is self-explanatory; both $Z[\pi_0(\Omega BM)]$ and $H_*(M)$ map to $H_*(\Omega BM)$, which is a ring under the Pontryagin product, and the map

$$Z[\pi_0(\Omega BM)] \otimes H_*(M) \longrightarrow H_*(\Omega BM)$$

is middle-linear over $Z[\pi_0(M)]$.

It remains to explain the limit in the first form of the statement. We have a map

$$H_*(M_\beta) \longrightarrow H_*(M_{\alpha\beta})$$

defined by left translation, that is, by left multiplication with $\alpha \in \pi_0(M)$. We have a diagram

$$\begin{array}{ccc} H_*(M_\beta) & \longrightarrow & H_*((\Omega BM)_{i_*\beta}) \\ \downarrow & & \downarrow \\ H_*(M_{\alpha\beta}) & \longrightarrow & H_*((\Omega BM)_{i_*\alpha\beta}) \, . \end{array}$$

So if this system is suitable for taking limits at all, we can take

$$\varinjlim_\beta H_*(M_\beta) \longrightarrow \varinjlim_\beta H_*((\Omega BM)_{i_*\beta}) \, ;$$

but on the right-hand side we have a system of isomorphisms, so the map from $H_*((\Omega BM)_0)$ to the limit is iso.

Well, we had better throw in enough assumptions so that our system is suitable for taking limits. We make the following assumptions.

(i) For all α, β in $\pi_0(M)$ there exist γ, δ in $\pi_0(M)$ such that

$$\gamma\alpha = \delta\beta \, .$$

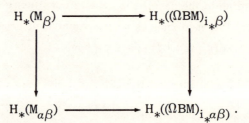

(ii) If in $\pi_0(M)$ we have

$$\alpha\gamma = \beta\gamma$$

then there exists δ in $\pi_0(M)$ such that

$$\delta\alpha = \delta\beta .$$

Both assumptions are trivial if $\pi_0(M)$ is commutative; in (i) one takes $\gamma = \beta$, $\delta = \alpha$ and in (ii) one takes $\delta = \gamma$. Now, in our examples (a), (b) above – and in all similar applications – the product map $\mu\colon M \times M \to M$ is actually homotopy-commutative. (Both examples come from suitable categories; the proof of homotopy-commutativity uses the natural isomorphism $c\colon \square \cong \square\tau$ but not its coherence.) So we are very happy to make any assumption weaker than homotopy-commutativity—in particular, that $\pi_0(M)$ is commutative. On the other hand, (i) and (ii) could be true without $\pi_0(M)$ being commutative.

These assumptions imply that $Z[\pi_0(\Omega BM)]$ is flat over $Z[\pi_0(M)]$. For example, in the case $\pi_0(M) = \{0,1,2,\cdots\}$, the ring of finite Laurent series $Z[t,t^{-1}]$ is flat over the polynomial ring $Z[t]$. The reason in the general case is the same as the reason in the special case; $Z[\pi_0(\Omega BM)]$ is a direct limit of free modules. They also imply that

$$Z[\pi_0(\Omega BM)] \otimes_{Z[\pi_0(M)]} H_*(M) \cong Z[\pi_0(\Omega BM)] \otimes_Z \varinjlim_\alpha H_*(M_\alpha) ,$$

so the two forms of the theorem are equivalent.

We need one further piece of data. Let $\gamma \in \pi_0(M)$. Then right translation by γ commutes with all the maps of the direct system, so it defines a map

$$\varinjlim_\alpha H_*(M_\alpha) \longrightarrow \varinjlim_\alpha H_*(M_\alpha) .$$

I would like this map to be iso for all $\gamma \in \pi_0(M)$. This is certainly true if M is homotopy-commutative, for then right translation by γ is homotopic to left translation by γ. It is also a necessary condition for the theorem to be true at all, because right translation by $i\gamma$ certainly gives an isomorphism

$$H_*(\Omega BM) \longrightarrow H_*(\Omega BM) .$$

With these assumptions one can give the following sketch proof of the theorem. On the right we have a fibering

$$\Omega BM \to EBM \to BM$$

and this fibering has a classical Serre spectral sequence. Here ΩBM acts on the left of EBM, and therefore the spectral sequence is a spectral sequence of left modules over $H_*(\Omega BM)$. The spectral sequence has the form

$$H_*(BM; S) \Rightarrow H_*(pt)$$

where S is a system of local coefficients, isomorphic over each point to $H_*(\Omega BM)$.

On the left we have a construction

$$M \to EM \to BM ,$$

and here M acts on EM over the identity map of BM. This construction need not be a fibering in any of the usual senses. However, it has a spectral sequence

$$H_*(BM; S') \Rightarrow H_*(pt) .$$

Here S' is a system of coefficients which is not even local, because the boundary maps which arise over the faces of simplexes in BM are not isomorphisms. However the coefficients at each point are isomorphic to $H_*(M)$. Moreover, this is a spectral sequence of left modules over $H_*(M)$.

I will assume without discussion that the technicians can define a comparison map from the second spectral sequence to the first, in which all the maps are maps of left modules over $H_*(M)$.

We may now localize the second spectral sequence by applying

$$Z[\pi_0(\Omega BM)] \otimes_{Z[\pi_0(M)]}$$

This functor preserves exactness, as we have said, and therefore the result is still a spectral sequence. Also the result still admits a comparison map to the first spectral sequence. But the boundary maps in the coefficient system S' are defined by right translation, so after localization they become iso by assumption, and after localization we get a local system of coefficients S'' which at each point is

$$Z[\pi_0(\Omega BM)] \otimes_{Z[\pi_0(M)]} H_*(M)$$

and maps to S.

Now we turn the handle of the comparison theorem for spectral sequences. Suppose as an inductive hypothesis that

$$Z[\pi_0(\Omega BM)] \otimes_{Z[\pi_0(M)]} H_*(M) \longrightarrow H_*(\Omega BM)$$

is iso in dimensions $< n$, so that $S'' \to S$ is iso in dimensions $< n$. Then we find that

$$H_0(BM; S''_n) \longrightarrow H_0(BM; S_n)$$

is iso, that is

$$Z \otimes_{Z[\pi_0(\Omega BM)]} Z[\pi_0(\Omega BM)] \otimes_{Z[\pi_0(M)]} H_n(M) \longrightarrow Z \otimes_{Z[\pi_0(\Omega BM)]} H_n(\Omega BM)$$

is iso, that is

$$\varinjlim_{\alpha} H_n(M_\alpha) \longrightarrow H_n((\Omega BM)_0)$$

is iso. This implies the inductive hypothesis in dimension n, which completes the sketch proof.

CHAPTER 4

TRANSFER

§4.1. *Introduction*

The notion of transfer is intimately bound up with the questions raised in Chapter 1. More precisely, it provides a piece of structure in each group of a cohomology theory – say for example in

$$E^0(X) = [X \cup (pt), \Omega^\infty E] -$$

which is not present in $[X \cup (pt), Y]$ in general; and this extra piece of structure reflects the infinite loop structure on $\Omega^\infty E$.

Any student of algebraic topology can be recommended to know at least a little about transfer, for this extra piece of structure provides one more weapon for carrying out our calculations and proofs. We need it in Chapter 5, which includes an application of the transfer; and it becomes essential in Chapter 6. It is for all these reasons that I now devote a short chapter to the transfer.

In the present section, §4.1, I try to set out the basic ideas in quasi-historical order. In §4.2 I will present the relationship between transfer and "structure maps" of the sort considered in Chapter 2. In §4.3 I will mention some of the formal properties of transfer which are the most fundamental in applications; this section ends with a kind of worked example.

The most classical case of transfer is for ordinary homology and cohomology. The treatment I shall indicate goes back to Eckmann [56]. Let

$$p : X \to Y$$

be an n-sheeted covering, that is, a locally trivally trivial fibering whose fibre is the discrete space with n points; we may also view it as a fibre-bundle in the sense of Steenrod, the structure group being the symmetric group Σ_n.

In order to define a transfer map

$$p^!: C_m(Y) \longrightarrow C_m(X) ,$$

it is sufficient to say that it is to be linear and give it on the generators. If we are using singular chains, the generators for $C_m(Y)$ are the singular simplexes $f: \sigma^m \to Y$. For each such singular simplex $f: \sigma^m \to Y$ there are precisely n singular simplexes $g: \sigma^m \to X$ such that $pg = f$. We set

$$p^! f = \Sigma g \mid pg = f ,$$

so that $p^! f$ is the sum of the singular simplexes over f.

We have

$$p^! d = dp^! ;$$

the transfer map is a chain map. In fact, if $d_i: \sigma^{m-1} \to \sigma^m$ are the face maps, and if g runs over the n singular simplexes such that $pg = f$, then gd_i runs over the n singular simplexes $h: \sigma^{m-1} \to X$ such that $ph = fd_i$.

Next suppose that B is a subspace of Y and $A = p^{-1}B \subset X$. Then $p^!$ maps $C_m(B)$ into $C_m(A)$. We obtain a chain map

$$p^!: C_m(Y,B) \longrightarrow C_m(X,A) .$$

Introducing coefficients in an abelian group Λ, we get a chain map

$$p^!: C_m(Y,B;\Lambda) \longrightarrow C_m(X,A;\Lambda)$$

and a cochain map

$$p_!: C^m(Y,B;\Lambda) \longleftarrow C^m(X,A;\Lambda) .$$

Finally, passing to homology and cohomology groups we get induced homomorphisms

$$p^!: H_m(Y,B;\Lambda) \longrightarrow H_m(X,A;\Lambda)\ ,$$

$$p_!: H^m(Y,B;\Lambda) \longleftarrow H^m(X,A;\Lambda)\ .$$

Note that these run in the opposite direction from the usual homomorphisms p_*, p^* induced by the projection

$$p: X,A \longrightarrow Y,B\ .$$

Both $p^!$ and $p_!$ are called "transfer"; they may be written Tr if there is no need to display p. The formal properties of $p^!$ and $p_!$ will receive more attention later.

It may be good to give a typical example of the use of the transfer; I give one which concerns the cohomology of groups. Let G be a finite group and S a subgroup of G; let BG be the classifying space of G, which is also an Eilenberg-MacLane space of type $(G,1)$; then a suitable finite covering of BG will be a space BS. This covering is an n-sheeted covering with $n = |G|/|S|$, the index of S in G. We have the following commutative diagram.

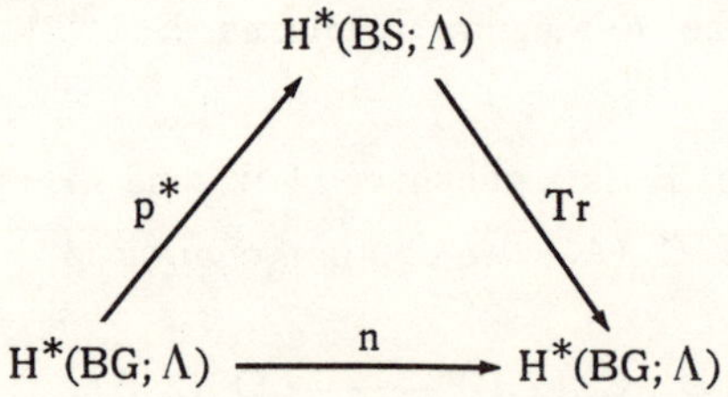

We see that if $n: \Lambda \to \Lambda$ is iso, then

$$p^*: H^*(BG;\Lambda) \longrightarrow H^*(BS;\Lambda)$$

is split mono. Some indications of the early work of J. Tate in this field may be found in [51] Chapter 12.

The next idea is that it may be possible to define transfers in cohomology theories other than the ordinary ones. In 1961 we find Atiyah doing so for complex K-theory – [17] p. 26. Let $p: X \to Y$ be a finite covering, and assume (for simplicity) that X and Y are finite-dimensional. Let ξ be a complex vector-bundle over X; we can construct a complex vector-bundle $p_! \xi$ over Y as follows. The fibre $(p_! \xi)_y$ at a point $y \in Y$ is to be given by

$$(p_! \xi)_y = \bigoplus_{px = y} \xi_x ;$$

these fibres fit together in an obvious way to form a vector-bundle. Since $p_!$ preserves addition, we get a homomorphism

$$p_! : K(X) \to K(Y) .$$

This is the "Atiyah transfer."

We may use the case of complex K-theory to suggest that for generalized cohomology theories we cannot expect to have $\operatorname{Tr} p^* = n$ as a formal property of transfers. For this we take $p : X \to Y$ to be the universal n-sheeted covering over $B\Sigma_n$; and we test the formula on the element 1, the trivial line bundle over $Y = B\Sigma_n$; then $\operatorname{Tr} p^* 1$ is not the trivial n-plane bundle. (It is instead the bundle corresponding to the representation in which Σ_n acts on C^n by permuting the coordinates.)

Still, the fact that we don't have $\operatorname{Tr} p^* = n$ turns out to make the general theory more interesting, rather than less.

We may say more generally that if we have an explicit geometrical definition for a generalized cohomology theory, then it is often possible to base on it an explicit, ad hoc, geometrical definition of the transfer. Such ad hoc definitions are very convenient for calculation, and can be illuminating. However, if we also possess a general theory of transfers, then of course it requires proof that our ad hoc definition agrees with the general definition, and such a proof can only be based on the details of the particular case. For the case of the "Atiyah transfer," compare [74], Proposition 2.4, p. 984.

The first reference I know for transfers in general is the work of Boardman in 1966. In [35] Chapter V §6, he defines eight "forward" or "Gysin" homomorphisms, all of which he calls "transfer," and of which (d) most obviously applies to our case (except that Boardman tends to assume that X and Y are manifolds). This is a source probably still well worth mining for ideas.

The next work which I must mention is the unpublished thesis of F. W. Roush in 1971 [122]. Roush states that some of his results were found independently by D. Quillen. Here we find a very important geometrical observation. Let $p : X \to Y$ be a locally trivial fibering with finite fibres; there is no need to assume that they all have the same cardinality if Y is not connected. Let us suppose for convenience that Y is a CW-complex and X has the corresponding CW-structure; let B be a subcomplex of Y and $A = p^{-1}B$, as above.

CONSTRUCTION 4.1.1. Such data determine a map of spectra

$$p^!: \Sigma^\infty(Y/B) \longrightarrow \Sigma^\infty(X/A) .$$

We note that here the construction $(X,A) \mapsto X/A$ is to be interpreted as a functor from the category of pairs to the category of spaces with base-point; it takes the space A and identifies it to a new point which is provided as the base-point; in particular, if A is empty we get $X/\phi = X \cup (\text{point})$.

It is clear from such a map $p^!$ of spectra we obtain transfers in all generalized cohomology theories at once. In fact, if E^* is a generalized cohomology theory corresponding to a spectrum E, then we get the following map.

$$\begin{array}{rcl} E^0(X,A) & = & [\Sigma^\infty(X/A),\ E] \\ & & \Big\downarrow (p^!)^* \\ E^0(Y,B) & = & [\Sigma^\infty(Y/B),\ E] \end{array}$$

Similarly with $E^0(\ )$ replaced by $E^n(\ )$.

This transfer is natural for maps $E \to F$; in other words, if $f: E \to F$ is a map of spectra, then the following diagram is commutative.

$$\begin{array}{ccc} [\Sigma^\infty(X/A),\ E] & \xrightarrow{f_*} & [\Sigma^\infty(X/A),\ F] \\ \Big\downarrow{\scriptstyle (p^!)^*} & & \Big\downarrow{\scriptstyle (p^!)^*} \\ [\Sigma^\infty(Y/B),\ E] & \xrightarrow{f_*} & [\Sigma^\infty(Y/B),\ F]\ . \end{array}$$

Conversely, if we want a transfer defined for all generalized cohomology theories at once, and natural in this sense, then our only hope is to look for a map

$$p^!: \Sigma^\infty(Y/B) \longrightarrow \Sigma^\infty(X/A)$$

(by Yoneda's Lemma, i.e. by substituting $E = \Sigma^\infty(X/A)$).

The geometrical idea involved in construction (4.1.1) is simple and illuminating. First suppose that X and Y are finite-dimensional, say of dimension $\leq d$. Form the product $Y \times I^n$ of Y with the standard n-cube. By taking n sufficiently large we can embed the covering $p: X \to Y$ in the projection map

$$Y \times (\mathrm{Int}\ I^n) \xrightarrow{\pi_1} Y\ ;$$

that is, we can construct a commutative diagram of the following form, in which e embeds X in $Y \times (\mathrm{Int}\ I^n)$.

$$\begin{array}{ccc} X & \xrightarrow{e} & Y \times I^n \\ & {\scriptstyle p}\searrow \quad \swarrow{\scriptstyle \pi_1} & \\ & Y & \end{array}$$

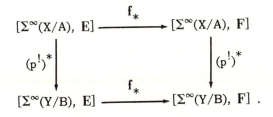

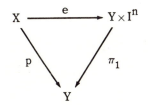

For this purpose we need enough space to make the sheets of the covering miss each other; $n > d$ will be sufficient. We can go on and construct a "cubular neighborhood" (cubical tubular neighborhood) of X in $Y \times I^n$. That is, we construct a commutative diagram of the following form.

$$\begin{array}{ccccc} X & \xrightarrow{e'} & X \times I^n & \xrightarrow{e''} & Y \times I^n \\ & {\scriptstyle p} \searrow & \downarrow {\scriptstyle p\pi_1} & \swarrow {\scriptstyle \pi_1} & \\ & & Y & & \end{array}$$

Here e' is the embedding of X in $X \times I^n$ using the centroid of I^n, while over each point $y \in Y$, e'' is the embedding of a finite number of non-overlapping little cubes in I^n. Each little cube $x \times I^n$ is embedded by a map

$$(x_1, x_2, \cdots, x_n) \longmapsto (m_1 x_1 + c_1,\ m_2 x_2 + c_2,\ \cdots,\ m_n x_n + c_n)$$

(see Chapter 2).

Of course, if you prefer to construct the embedding e'' of $X \times I^n$ directly without using the embedding e of X as scaffolding, that's fine too – the embedding e'' is all we need.

In any case, by collapsing to a point the outside of the cubular neighborhood, and also the subspace $B \times I^n$, we get a map

$$\begin{array}{ccc} \dfrac{Y \times I^n}{Y \times \dot{I}^{n-1} \cup B \times I^n} & \xrightarrow{p^!} & \dfrac{X \times I^n}{X \times \dot{I}^{n-1} \cup A \times I^n} \\ \updownarrow = & & \updownarrow = \\ \Sigma^n(Y/B) & & \Sigma^n(X/A)\,. \end{array}$$

This is the required map.

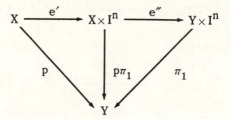

On the face of it the construction involves choice, and we must prove independence of the choice. Suppose the construction performed twice, with choices n_0, n_1 of n, and with choices e_0, e_1 of e and so on. There is no difficulty in increasing n_0 and n_1 to a common value n, and having n as large as we need; for by increasing I^{n_0} to I^n and making the last coordinates $x_{n_0+1}, \cdots, x_n$ play a passive role throughout we replace $p_0^!$ by its suspension $S^{n-n_0} p_0^!$, and similarly for $p_1^!$. After that there is no difficulty in constructing a diagram

$$\begin{array}{ccccc} I\times X & \xrightarrow{\;e'\;} & I\times X\times I^n & \xrightarrow{\;e''\;} & I\times Y\times I^n \\ & {\scriptstyle 1\times p}\searrow & \Big\downarrow {\scriptstyle 1\times p\pi_1} & \swarrow{\scriptstyle 1\times \pi_1} & \\ & & I\times Y & & \end{array}$$

which agrees with the given diagrams at the ends $t = 0$, $t = 1$ of the time-interval I. This leads to a map

$$\frac{I\times \Sigma^n(Y/B)}{I\times pt} \longrightarrow \frac{I\times \Sigma^n(X/A)}{I\times pt},$$

and this provides the required homotopy.

Finally, if X and Y are infinite-dimensional complexes we have to proceed in this way for the skeletons X^d, Y^d; as d increases we increase n correspondingly, and we arrange an obvious compatibility between what we do for d and what we do for $d+1$. It is for this purpose that we really have to go to the limit and use the spectra $\Sigma^\infty(X/A)$, $\Sigma^\infty(Y/B)$.

We ask the reader to satisfy himself that when he applies ordinary homology or cohomology to the map

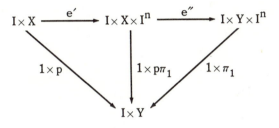

$$p^!: \Sigma^\infty(Y/B) \longrightarrow \Sigma^\infty(X/A)$$

of (4.1.1), he gets a result consistent with the classical construction of Eckmann with which we started. (Replace the singular simplexes by cells; or see [74] Proposition 2.1 p. 983.)

The theory of Roush also provides a relation between the transfer and "structure maps" of the sort considered in Chapter 2; we will say something about this in §4.2.

It is clear that the maps $p^!$ of Construction 4.1.1 must have good formal properties. For the moment we defer these, but we will say something about them in §4.3.

Next I interpose a comment on our main theme. We have seen that before a contravariant functor $k(X)$ can be the 0th term of a cohomology theory, it is certainly necessary that it should possess transfer maps

$$p_! : k(X) \to k(Y)$$

for all finite covering maps $p : X \to Y$; and no doubt these should satisfy certain axioms if one wished to make such axioms explicit. At one time there was even a conjecture – [141] p. 50–that this was a sufficient condition; that is, any representable group-valued functor which comes equipped with transfer maps satisfying suitable conditions is the 0th term of a cohomology theory. However, it is now known that this is false; a counterexample was announced by D. Kraines at the conference in Evanston, 1977.

We thus believe that in problems about infinite loop spaces, transfer conditions are necessary but in general not sufficient. This leaves open the possibility that they may be sufficient in particular cases of interest; we shall return to this theme in Chapter 6.

Transfer came to the attention of the general topological public when Kahn and Priddy published their well-known paper in 1972 [74]. Kahn and Priddy wrote: "the existence of the transfer seems to be well known, but we know of no published account''. The topological world thus learnt that all well-informed persons were supposed to know about transfer, although hardly anyone did unless they were lucky.

It may perhaps be worth stating the result of Kahn and Priddy, for the rapid spread of a general conviction that the transfer was very good business owed much to the fact that they solved a problem of some standing in homotopy-theory.

Take

$$\Omega^\infty \Sigma^\infty S^0 = \lim_{n \to \infty} \Omega^n S^n ;$$

take one component of it, say $(\Omega^\infty \Sigma^\infty S^0)^0$; and localize it at the prime 2, getting say $(\Omega^\infty \Sigma^\infty S^0)^0_{(2)}$. Then this space occurs as a direct factor of $\Omega^\infty \Sigma^\infty RP^\infty$; more precisely, we have

$$\Omega^\infty \Sigma^\infty RP^\infty = (\Omega^\infty \Sigma^\infty S^0)^0_{(2)} \times (?) ,$$

where the ''$\times$'' means Cartesian product and the ''(?)'' means an unknown factor. In particular, the stable homotopy groups of RP^∞ map onto the 2-primary components of the stable homotopy groups of spheres in positive degrees, and they map onto by a split epimorphism. The epimorphism is induced by a well-known map of spectra. However, the homomorphism in the reverse direction, which provides the splitting, cannot be induced by a map of spectra; it can only be induced by a map of spaces. The result is therefore interesting from a methodological point of view, for it seems to involve essential use of ''unstable'' geometry to prove a ''stable'' result.

There is a corresponding result when the prime 2 is replaced by the prime p. One can replace RP^∞ by BZ_p, but that is unnecessarily big; one can replace RP^∞ by $B\Sigma_p$, but that is not p-primary; it is perhaps best to use

$$(B\Sigma_p)^+ \, (p) \cdot$$

The next work on which I must comment is that of Becker and Gottlieb in 1975 [33]. Becker and Gottlieb treat a fibering more general than a covering. More precisely, they suppose given a fibre bundle $p: E \to B$ whose fibre F is a compact smooth manifold, whose structural group G is a compact Lie group acting smoothly on F, and whose base B is a finite complex. This situation is strikingly similar to Boardman's transfer (h), [35], pp. 45-48. Then Becker and Gottlieb construct a map

$$p^!: \Sigma^\infty(B/\emptyset) \longrightarrow \Sigma^\infty(E/\emptyset)$$

which can be used to induce transfer homomorphisms. Note that these transfer homomorphisms preserve degrees; that is, they give maps

$$(p^!)^*: H^n(E) \longrightarrow H^n(B)\ ;$$

there is no change in the degrees as you get with some Gysin homomorphisms. In fact, in ordinary cohomology the Becker-Gottlieb construction gives a result which can be explained in terms which were already known. For this purpose I must recall some material. Suppose that the manifold F is of dimension d; then by considering the terms $E^{p,q}$ of the spectral sequence with $q = d$, we obtain a map

$$H^n(E) \longrightarrow H^{n-d}(B; H^d(F))\ ;$$

using the fundamental class in the fibre, we obtain (under appropriate hypotheses) a map

$$H^n(E) \longrightarrow H^{n-d}(B)\ .$$

This map is called "integration over the fibres" [41] because that name is appropriate in the special case of de Rham cohomology. Then in ordinary cohomology the Becker-Gottlieb construction gives the composite of (i) multiplication with the Euler class of the tangent bundle along the fibres, and (ii) "integration over the fibres."

This result allows Becker and Gottlieb to carry over to Lie groups G results and methods which were previously available for finite groups. Let us consider an example, which is modelled on the result for finite groups given above.

PRETHEOREM 4.1.2. *Let* G *be a compact connected Lie group, let* N *be the normalizer of a maximal torus in* G, *and let* E^* *be any generalized cohomology theory. Then the map*

$$E^*(BG) \longrightarrow E^*(BN)$$

(*induced by the inclusion of* N *in* G) *is split mono.*

Sketch proof. Consider the fibering

$$G/N \longrightarrow BN \xrightarrow{p} BG .$$

In this fibering, G/N is a compact smooth manifold, as required; moreover the Euler-Poincaré characteristic $\chi(G/N)$ is 1. If we were working in ordinary cohomology, this would allow us to prove that the composite

$$H^*(BG) \xrightarrow{p^*} H^*(BN) \xrightarrow{Tr} H^*(BG)$$

is 1. In generalized cohomology we cannot conclude that this composite is 1. However, this composite should be induced by a map of spectra, say

$$f\colon \Sigma^\infty(BG/\emptyset) \longrightarrow \Sigma^\infty(BG/\emptyset)$$

which induces the identity map in ordinary homology and cohomology; then f should be an equivalence, and this yields the required splitting.

Next I must explain why I have called (4.1.2) a "pretheorem." The sketch proof given is not complete, because as I have said Becker and Gottlieb prefer to construct their transfer only when the base is a finite complex, and BG is not a finite complex. For this reason, Becker and Gottlieb do not state their result exactly in the form given as (4.1.2); they state instead a "finite-complex analogue" of (4.1.2). One can in fact use the result of Becker and Gottlieb to deduce the most important corollaries which should flow from (4.1.2); for in each case one adds technical arguments to show that it is sufficient to consider finite complexes. In some cases one has such technical arguments anyway (for example, to make BG approximate BU or BO) and then the added nuisance is slight.

However, it eases our thinking if we have results in their simplest and most general form; and from this point of view one might prefer a different approach, as follows. The sketch proof above mentions a map of spectra; however Becker and Gottlieb prefer not to use spectra; indeed it seems that they assume their base is finite mainly in order to avoid using spectra. It seems that by using spectra, one might avoid the restriction to finite bases. It would probably be desirable to have this worked out. If it can be done, then (4.1.2) is a theorem as it stands, by the proof sketched.

At all events, we conclude that in principle, Becker-Gottlieb transfer should permit one to obtain results about Lie groups comparable with those which classical transfer yields about finite groups.

In subsequent work the notion of Becker-Gottlieb transfer has been generalized, and the conditions needed for constructing it have been weakened [31].

There is a nice short summary of a construction of Becker-Gottlieb transfer, with applications, in [79].

§4.2. *Transfer and structure maps*

In this section our first object is to relate transfer to the sort of structure map considered in Chapter 2. Once we have got a theory in those terms, our second task is to relate it to the approach given in §4.1 – meaning Construction 4.1.1. Finally we consider how certain good properties of the transfer correspond to good properties of structure maps.

Our first aim is to give a (1-1) correspondence between "transfers" and homotopy classes of "structure maps." Of course we must begin by defining these terms.

A "transfer" will assign to each n-sheeted covering $p : X \to A$ a function

$$p_! : [X,V] \longrightarrow [A,W] .$$

Here n is to be a fixed number; we are dealing with transfers defined for n-sheeted coverings. V and W are to be fixed spaces; we think of $[\ ,V]$ as analogous to one cohomology theory (or rather one cohomology group), and $[\ ,W]$ as analogous to another. $[X,V]$ means the set of homotopy classes of maps from X to V, where neither maps nor homotopies are required to preserve any base-points; similarly for $[A,W]$. (We choose this approach because we want to apply the functor $[\ ,V]$ directly to our covering space X and not to $X \cup (pt)$.) To cater for spaces V and W which may be different is not just excess generality; it is done for good and sufficient reason, as will appear. As for X and A, we depart from the notation in §4.1 because in this section we are content to deal with the absolute case and forget the relative case.

Our functions $p_!$ are required to satisfy one axiom, and it is a naturality axiom. To state it, we need to define maps. A map of n-sheeted coverings should be a pullback diagram

$$\begin{array}{ccc} X & \xrightarrow{\ \xi\ } & Y \\ {\scriptstyle p}\big\downarrow & & \big\downarrow{\scriptstyle q} \\ A & \xrightarrow{\ \alpha\ } & B \end{array}$$

so that ξ maps each fiber of p bijectively to a fiber of q. We demand that for each such map of n-sheeted coverings the following diagram should be commutative.

$$\begin{array}{ccc} [X,V] & \xleftarrow{\ \xi^*\ } & [Y,V] \\ {\scriptstyle p_!}\downarrow & & \downarrow{\scriptstyle q_!} \\ [A,W] & \xleftarrow{\ \alpha^*\ } & [B,W] \end{array}$$

This completes our definition of a "transfer."

A "structure map" is to be a map

$$\theta\colon (E\Sigma_n \times V^n) \times_{\Sigma_n} (pt) \longrightarrow W\ .$$

I apologize for spelling out the details, but I am going to use them. Here the group "Σ_n" is the symmetric group on the n letters $\{1,2,\cdots,n\}$. The space "$E\Sigma_n$" is the total space of a universal Σ_n-bundle; thus it is contractible, and Σ_n acts freely on its right. The group Σ_n also acts on the right of the Cartesian product V^n, as in §1.3. The group Σ_n acts on the right of the product $E\Sigma_n \times V^n$ by acting on both factors. It acts trivially on the left of (pt). The product $X \times_G Y$ over a group G is defined as a quotient of $X \times Y$ in which we make the identification

$$(xg,y) \simeq (x,gy)\ .$$

It may seem that $X \times_G (pt)$ is unnecessarily cumbersome notation for X/G; at this point it is, but later we need to relate this space to other spaces of the form $X \times_G Y$ with Y non-trivial.

What we have just given is the choice of details which seems convenient in this section; but it is easy to reconcile it with the different choices made by other authors (or by the present author on a different page). For example, we may change the action of G from one side of the space Y to the other by the rule

$$gy = yg^{-1}\ ;$$

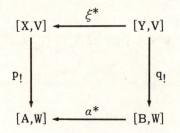

and using this remark as necessary, we may replace

$$(E\Sigma_n \times V^n) \times_{\Sigma_n} (pt)$$

by

$$E\Sigma_n \times_{\Sigma_n} V^n , \qquad V^n \times_{\Sigma_n} E\Sigma_n$$

or whatever else is needed.

This completes our account of "structure maps." Structure maps are classified under homotopy, where neither maps nor homotopies are required to preserve any base-points.

It is not enough just to assert the existence of a (1-1) correspondence between "transfers" and "structure maps"; we need to know what the (1-1) correspondence is. In particular, given a structure map θ, we need to know how to construct the corresponding transfer.

To explain this conveniently we first recall how to construct the principal bundle associated to an n-sheeted covering $p: X \to A$. We define a subset $\overline{X} \subset X^n$, legislating that a function

$$\underline{x}: \{1,2,\cdots,n\} \longrightarrow X$$

shall lie in $\overline{X}$ if and only if it is a bijection from $\{1,2,\cdots,n\}$ to some fiber of p. Thus Σ_n acts freely on the right of $\overline{X}$, and we have an obvious identification

$$\overline{X} \times_{\Sigma_n} (pt) \cong A .$$

Suppose then that we are given a structure map

$$\theta: (E\Sigma_n \times V^n) \times_{\Sigma_n} (pt) \longrightarrow W$$

an n-sheeted covering $p: X \to A$ and a map $u: X \to V$. We can construct a Σ_n-map

$$\lambda: \bar{X} \longrightarrow E\Sigma_n$$

because $E\Sigma_n$ is universal. We can construct a Σ_n-map

$$\mu: \bar{X} \to V^n$$

by taking

$$\bar{X} \xrightarrow{i} X^n \xrightarrow{u^n} V^n$$

where i is the inclusion. With λ and μ as components we can form a Σ_n-map

$$\bar{X} \xrightarrow{(\lambda,\mu)} E\Sigma_n \times V^n .$$

We define $p_!u$ to be the following composite.

$$A \cong \bar{X} \times_{\Sigma_n} (pt) \xrightarrow{(\lambda,\mu)\times_{\Sigma_n} 1} (E\Sigma_n \times V^n) \times_{\Sigma_n} (pt) \xrightarrow{\theta} W .$$

We have to do a little checking to see that this gives a well-defined function

$$p_!: [X,V] \longrightarrow [A,W] .$$

Any two choices of λ differ by a Σ_n-homotopy since $E\Sigma_n$ is universal; and similarly if we alter u by a homotopy we alter μ by a Σ_n-homotopy; so these changes alter $(\lambda,\mu)\times_{\Sigma_n} 1$ by a homotopy. This completes the construction of $p_!$ from θ.

Some authors seem to believe that this construction of $p_!$ calls for the use of coset representatives. If anyone still feels that coset representatives help to reveal the map

$$\bar{X} \xrightarrow{i} X^n \xrightarrow{u^n} V^n$$

in its full blinding simplicity, I have an opinion of his taste.

THEOREM 4.2.1 (after Roush [122]). *The construction above gives a (1-1) correspondence from homotopy classes of structure maps*

$$\theta: (E\Sigma_n \times V^n) \times_{\Sigma_n} (pt) \longrightarrow W$$

to "transfers" of the sort described.

In order to prove such a result, the obvious method is the method of the "universal example", that is, Yoneda's lemma. This becomes most conceptual if we fix up a suitable category $\mathcal{C}$ to work in. Let us legislate that an *object* of our category is to be a pair, in which the first element is an n-sheeted covering $p: X \to A$ and the second element is a class $u \in [X,V]$. A *morphism* in our category $\mathcal{C}$, from $(p: X \to A;\ u)$ to $(q: Y \to B;\ v)$ is to be a homotopy class of maps

$$\begin{array}{ccc} X & \xrightarrow{\ \xi\ } & Y \\ {\scriptstyle p}\downarrow & & \downarrow{\scriptstyle q} \\ A & \xrightarrow{\ \alpha\ } & B \end{array}$$

such that $\xi^* v = u$. Here a "homotopy" is to be a map of n-sheeted coverings

$$\begin{array}{ccc} I \times X & \xrightarrow{\ \eta\ } & Y \\ {\scriptstyle 1 \times p}\downarrow & & \downarrow{\scriptstyle q} \\ I \times A & \xrightarrow{\ \beta\ } & B\,. \end{array}$$

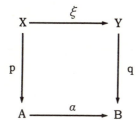

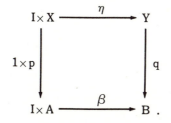

We will actually prove that this category $\mathcal{C}$ has a terminal object $(q: Y \to B;\ v)$, that is, an object which receives exactly one morphism from any other object. Once this is proved, it certainly follows that there is a (1-1) correspondence between transfers and classes in [B,W]. Our construction of the terminal object will show that

$$B = (E\Sigma_n \times V^n) \times_{\Sigma_n} (pt) ,$$

so that a map $B \to W$ is a structure map of the sort considered. Similarly, our proof that the terminal object receives only one morphism will lead to the description of the (1-1) correspondence which has been given above. In fact, let

$$\begin{array}{ccc} X & \xrightarrow{\ \xi\ } & Y \\ {\scriptstyle p}\downarrow & & \downarrow{\scriptstyle q} \\ A & \xrightarrow{\ a\ } & B \end{array}$$

be the unique morphism in C whose existence is asserted; then on general grounds, given a structure map $\theta : B \to W$, the corresponding transfer must have

$$p_! u = \theta a ;$$

so all we need to see is the construction of a.

Before proceeding, it may be useful to run over the relation between an n-sheeted covering and the associated principal bundle. Given an n-sheeted covering $p : X \to A$, we have constructed $\overline{X}$. Conversely, given a space E and a free action of Σ_n on the right of E, we can construct an n-sheeted covering by

$$\begin{array}{c} E \times_{\Sigma_n} \{1,2,\cdots,n\} \\ \Big\downarrow {\scriptstyle 1 \times_{\Sigma_n} c} \\ E \times_{\Sigma_n} (pt) . \end{array}$$

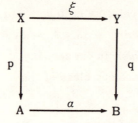

(The map $c\colon \{1,2,\cdots,n\} \longrightarrow (pt)$ is the unique one, called c for "constant.")

These two processes are inverse. The detail which we need to know is that if we start from $p: X \to A$, we have the following commutative diagram.

$$\begin{array}{ccc} \bar{X}\times_{\Sigma_n}\{1,2,\cdots,n\} & \xrightarrow{\ \cong\ } & X \\ \Big\downarrow{\scriptstyle 1\times_{\Sigma_n} c} & & \Big\downarrow{\scriptstyle p} \\ \bar{X}\times_{\Sigma_n}(pt) & \xrightarrow{\ \cong\ } & A \end{array}$$

Here the upper arrow is given by

$$\begin{array}{ccc} \bar{X}\times_{\Sigma_n}\{1,2,\cdots,n\} & & \\ \Big\downarrow{\scriptstyle i\times_{\Sigma_n} 1} & & \\ X^n\times_{\Sigma_n}\{1,2,\cdots,n\} & \xrightarrow{\ ev\ } & X \end{array}$$

where the evaluation map ev is defined by

$$ev(\underline{x}, i) = \underline{x}(i) .$$

We also need to know that there is a (1-1) correspondence between maps

$$\begin{array}{ccc} X & \xrightarrow{\ \xi\ } & Y \\ {\scriptstyle p}\Big\downarrow & & \Big\downarrow{\scriptstyle q} \\ A & \xrightarrow{\ \alpha\ } & B \end{array}$$

of n-sheeted coverings, and Σ_n-maps

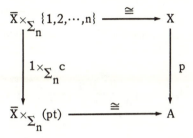

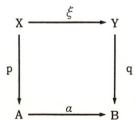

$$\bar{\xi}: \bar{X} \to \bar{Y}$$

of the associated principal bundles. This (1-1) correspondence is valid at the level of maps, before we pass to homotopy classes.

We proceed to construct the proposed terminal object. The n-sheeted covering is to be that associated to the principal bundle $E\Sigma_n \times V^n$; explicitly, it is as follows.

$$\begin{array}{c} Y = (E\Sigma_n \times V^n) \times_{\Sigma_n} \{1,2,\cdots,n\} \\ \Big\downarrow {\scriptstyle 1 \times_{\Sigma_n} c} \\ B = (E\Sigma_n \times V^n) \times_{\Sigma_n} (pt)\ . \end{array}$$

The map $v: Y \to V$ is to be the following composite.

$$\begin{array}{l} Y = (E\Sigma_n \times V^n) \times_{\Sigma_n} \{1,2,\cdots,n\} \\ \qquad\qquad \Big\downarrow {\scriptstyle \pi_2 \times_{\Sigma_n} 1} \\ \qquad\quad V^n \times_{\Sigma_n} \{1,2,\cdots,n\} \xrightarrow{\ ev\ } V\ . \end{array}$$

Here the map

$$\pi_2: E\Sigma_n \times V^n \longrightarrow V^n$$

is projection onto the second functor, and ev is defined by evaluation, as before.

We recall that we have already constructed a map

$$\bar{X} \xrightarrow{(\lambda, \mu)} E\Sigma_n \times V^n$$

of principal bundles; we can now use it to induce a map of n-sheeted coverings; the result is the following.

$$
\begin{array}{ccc}
X \cong \bar{X}\times_{\Sigma_n}\{1,2,\cdots,n\} & \xrightarrow{(\lambda,\mu)\times_{\Sigma_n}1} & (E\Sigma_n\times V^n)\times_{\Sigma_n}\{1,2,\cdots,n\} = Y \\
p\downarrow \quad \downarrow 1\times_{\Sigma_n}c & & \downarrow 1\times_{\Sigma_n}c \\
A \cong \bar{X}\times_{\Sigma_n}(pt) & \xrightarrow{(\lambda,\mu)\times_{\Sigma_n}1} & (E\Sigma_n\times V^n)\times_{\Sigma_n}(pt) = B\,.
\end{array}
$$

Let us abbreviate this diagram as

$$
\begin{array}{ccc}
X & \xrightarrow{\xi} & Y \\
p\downarrow & & \downarrow q \\
A & \xrightarrow{\alpha} & B
\end{array}
$$

where $\xi = \xi(\lambda,\mu)$, $\alpha = \alpha(\lambda,\mu)$.

LEMMA 4.2.2. (i) *For each map* $u : \bar{X} \to V$ *the formula* $\mu = u^n i$ *gives the unique* Σ_n*-map* $\mu : X \to V^n$ *such that the resulting diagram*

$$
\begin{array}{ccc}
 & Y & \\
\xi=\xi(\lambda,\mu) \swarrow & & \searrow v \\
X & \xrightarrow{u} & V
\end{array}
$$

is commutative (strictly — not just up to homotopy).

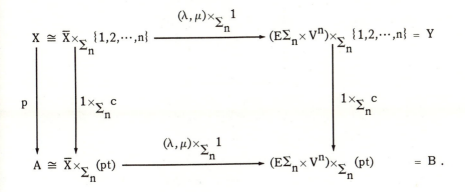

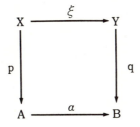

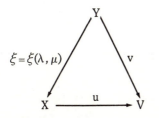

(ii) *The diagram*

$$\begin{array}{ccc} X & \xrightarrow{\xi} & Y \\ {\scriptstyle p}\downarrow & & \downarrow{\scriptstyle q} \\ A & \xrightarrow{\alpha} & B \end{array}$$

constructed above represents the unique morphism in $\mathcal{C}$ *from the object* $(p\colon X \to A;\ u)$ *to the object* $(q\colon Y \to B;\ v)$.

Proof. Part (i) is elementary and simply involves checking from the definitions.

To prove part (ii), suppose given two maps of coverings, say

$$\begin{array}{ccc} X & \xrightarrow{\xi_0} & Y \\ {\scriptstyle p}\downarrow & & \downarrow{\scriptstyle q} \\ A & \xrightarrow{\alpha_0} & B \end{array} \qquad\qquad \begin{array}{ccc} X & \xrightarrow{\xi_1} & Y \\ {\scriptstyle p}\downarrow & & \downarrow{\scriptstyle q} \\ A & \xrightarrow{\alpha_1} & B \end{array}$$

where $v\xi_0 \simeq v\xi_1$. These two maps of coverings correspond to Σ_n-maps (λ_t, μ_t) for $t = 0,1$. Here λ_0 and λ_1 extend to a Σ_n-map

$$\lambda\colon I \times \overline{X} \longrightarrow E\Sigma_n$$

because $E\Sigma_n$ is universal. We want to see that μ_0, μ_1 extend to a Σ_n-map

$$\mu\colon I \times \overline{X} \longrightarrow V^n\,;$$

and for this purpose it is sufficient to apply part (i) with u replaced by a homotopy

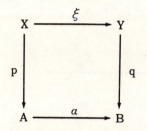

$$I \times X \to V$$

from $v\xi_0$ to $v\xi_1$. This shows that our two original maps of n-sheeted coverings were homotopic; and this proves Lemma 4.2.2.

Theorem 4.2.1 follows immediately from Lemma 4.2.2, according to the discussion above.

Kahn and Priddy give the name "pretransfer" to the special case in which the structure map θ is the identity map 1, so that

$$W = (E\Sigma_n \times V^n) \times_{\Sigma_n} (pt) .$$

From the account above it is clear that it is a transfer; that it is initial among transfers; and that it needs no other comment.

THEOREM 4.2.3. *If in the construction above we insert the structure map*

$$(E\Sigma_n \times V^n) \times_{\Sigma_n} (pt) \longrightarrow V$$

which any infinite loop space $V = \Omega^\infty V$ *has, then the resulting transfer agrees with Construction 4.1.1.*

This result may be found in Kahn and Priddy – [74] Proposition 1.7.

To effect the required reconciliation, we take $E\Sigma_n$ to be the space $P_{1,n}$ of "n little cubes in a big cube" indicated in Chapter 2. (One can get the idea of the proof by considering r-dimensional cubes I^r for a fixed value of r; but the space of n little r-cubes is not strictly a space $E\Sigma_n$, and the proper proof involves compatibility as you vary r and a passage to the limit, as in §4.1.)

Let $\overline{x}$ be a point of $\overline{X}$, which we can interpret as n points $x_1, x_2, \cdots, x_n$ in X, given in order and forming one complete fibre of $p\colon X \to A$. Then clearly Construction 4.1.1 assigns to them n little cubes

$J_1, J_2, \cdots, J_n$ given in order in the standard cube I. This construction gives a map

$$\lambda: \overline{X} \longrightarrow P_{1,n},$$

and clearly λ is a Σ_n-map. (Conversely, a Σ_n-map

$$\lambda: \overline{X} \longrightarrow P_{1,n}$$

defines a geometrical embedding of the sort considered in §4.1, if one wants to proceed in that direction.) Now detailed checking from the definitions shows that the accounts in §4.1 and §4.2 correspond under adjointness.

In some circumstances we may obtain a structure map

$$(E\Sigma_n \times V^n) \times_{\Sigma_n} (pt) \longrightarrow V$$

by direct construction rather than by applying Theorem 4.2.3. For an easy example, let V be an Eilenberg-MacLane space of type (π, m); then V is an infinite loop space (see Chapter 1); but equally, we can construct the structure map. Since homotopy classes of maps into V are in (1-1) correspondence with cohomology classes, it is sufficient to discuss the cohomology group $H^m((E\Sigma_n \times V^n) \times_{\Sigma_n} (pt); \pi)$ and specify an appropriate class there. Alternatively, we may choose a model for V which is actually a commutative group, and consider the map

$$(E\Sigma_n \times V^n) \times_{\Sigma_n} (pt) \longrightarrow V$$

which carries $(e, v_1, v_2, \cdots, v_n)$ to the product $v_1 v_2 \cdots v_n$.

Even if our structure maps do not come from the source indicated in Theorem 4.2.3, the reader might expect to see how good properties of the transfer correspond to good properties of the structure maps. This can certainly be done. For example, let us suppose that V and W are

H-spaces, so that [,V] and [,W] are group-valued functors; we can then ask for a condition that the transfer corresponding to θ be a homomorphism of groups. It is clear that this must translate out into a homotopy-commutative diagram involving θ and the two H-space structure maps μ_V, μ_W. It is even clear how to obtain it. We have

$$[X, V\times V] = [X,V]\times[X,V]$$
$$[A, W\times W] = [A,W]\times[A,W] ,$$

and we can find a structure map Θ for which the corresponding transfer is

$$p_!\times p_!: [X,V]\times[X,V] \longrightarrow [A,W]\times[A,W] .$$

In fact, the necessary structure map Θ has to map into $W\times W$, so it needs two components; we define the first to be

$$\begin{array}{ccc}
(E\Sigma_n\times(V\times V)^n)\times_{\Sigma_n}(pt) & & \\
\Big\downarrow (1\times\pi_1^n)\times_{\Sigma_n} 1 & & \\
(E\Sigma_n\times V^n)\times_{\Sigma_n}(pt) & \xrightarrow{\ \theta\ } & W
\end{array}$$

and we define the second to be

$$\begin{array}{ccc}
(E\Sigma_n\times(V\times V)^n)\times_{\Sigma_n}(pt) & & \\
\Big\downarrow (1\times\pi_2^n)\times_{\Sigma_n} 1 & & \\
(E\Sigma_n\times V^n)\times_{\Sigma_n}(pt) & \xrightarrow{\ \theta\ } & W .
\end{array}$$

(Here of course $\pi_1 : V \times V \to V$ is projection onto the first factor and π_2 is projection onto the second factor.) Now the diagram we want is the following.

$$\begin{array}{ccc} (E\Sigma_n \times (V\times V)^n) \times_{\Sigma_n} (pt) & \xrightarrow{\Theta} & W \times W \\ \Big\downarrow {\scriptstyle (1\times \mu_V^n)\times_{\Sigma_n} 1} & & \Big\downarrow {\scriptstyle \mu_W} \\ (E\Sigma_n \times V^n) \times_{\Sigma_n} (pt) & \xrightarrow{\theta} & W \end{array}$$

If Θ is written out explicitly in terms of θ, this gives the required diagram, whose homotopy-commutativity is necessary and sufficient for the transfer corresponding to θ to be a homomorphism.

Evidently one can proceed in this way for other proposed good formal properties of the transfer. For example, suppose that $p : X \to Y$ is an m-sheeted covering and $q : Y \to Z$ is an n-sheeted covering; then the composite

$$X \xrightarrow{p} Y \xrightarrow{q} Z$$

is an mn-sheeted covering; we might like to have

$$(pq)_! = p_! q_!$$

for all such p, q. Clearly this approach must correspond to some diagram involving θ_n, θ_m and θ_{mn}. A similar situation arises when an (m+n)-sheeted covering splits up as the disjoint union of an m-sheeted covering and an n-sheeted covering.

The reason we do not proceed in this way is a lack of applications. In the interesting applications transfers come from infinite loop spaces; so on the one hand the transfers have all the good properties we want, and on the other the structure maps take part in all the diagrams that the most avid enthusiast could desire; we don't need to prove any equivalence.

This suggests, incidentally, that the diagrams corresponding to good properties of the transfer will all be known diagrams in the theory of infinite loop-spaces. The difference is that in the theory of transfer they will be asked to commute up to homotopy, while in the theory of infinite loop spaces they may be asked to commute either exactly, or up to an infinity of higher homotopies. This is the basic reason why the passage from "infinite-loop data" to "transfer conclusions" is not reversible.

§4.3. *Formal properties of the transfer*

In this section the transfer will always be as in §4.1. Thus the input is a covering

$$p : X,A \to Y,B$$

in which the fibres are finite and $A = p^{-1}B$, and we assume for convenience that our spaces are CW-complexes; while the output is a map of spectra

$$p^{!}: \Sigma^{\infty}(Y/B) \longrightarrow \Sigma^{\infty}(X/A)\ .$$

We will note some formal properties of this construction. The section ends with a sort of worked example, in which I suggest how one of these properties can be applied.

The first obvious property is naturality. Suppose given the following commutative diagram.

$$\begin{array}{ccc} X,A & \xrightarrow{\ \xi\ } & X',A' \\ {\scriptstyle p}\big\downarrow & & \big\downarrow{\scriptstyle p'} \\ Y,B & \xrightarrow{\ \eta\ } & Y',B' \end{array}$$

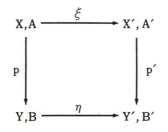

We assume as usual that p and p' are coverings with finite fibres, and $A = p^{-1}B$, $A' = (p')^{-1}B'$; also η maps B into B' (whence ξ maps A into A'); and finally, p is the pullback of p', that is, ξ maps each fibre of p bijectively to the corresponding fibre of p'. Then we conclude that the following diagram is commutative (up to homotopy).

(4.3.1)
$$\begin{array}{ccc}
\Sigma^\infty(X/A) & \xrightarrow{\Sigma^\infty \xi} & \Sigma^\infty(X'/A') \\
{\scriptstyle p^!}\downarrow & & \downarrow{\scriptstyle (p')^!} \\
\Sigma^\infty(Y/B) & \xrightarrow{\Sigma^\infty \eta} & \Sigma^\infty(Y'/B')
\end{array}$$

The verification is elementary.

The next property is used to show that after one applies a cohomology functor, transfer commutes with coboundary maps. Let $p: X,A \to Y,B$ be a covering with finite fibres and let $q: A \to B$ be its restriction to A. We can use the quotient map $X \to X/A$ to start the following cofibre sequence:

$$X \longrightarrow X/A \xrightarrow{j} (X/A) \cup CX \simeq SA ,$$

and similarly for $Y \to Y/B$. Then we conclude that the following diagram is commutative (up to homotopy).

(4.3.2)
$$\begin{array}{ccc}
\Sigma^\infty(X/A) & \xrightarrow{\Sigma^\infty j} & \Sigma^\infty SA \simeq S\Sigma^\infty A \\
{\scriptstyle p^!}\uparrow & & \uparrow{\scriptstyle S(q^!)} \\
\Sigma^\infty(Y/B) & \xrightarrow{\Sigma^\infty j} & \Sigma^\infty SB \simeq S\Sigma^\infty B
\end{array}$$

Sketch proof: one must carefully remember that the finite covering over Y does not extend over CY. Nevertheless, the fact that one can make the square

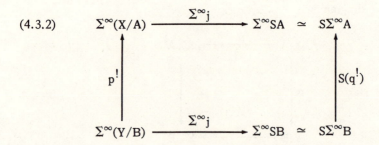

$$\begin{array}{ccc}
\Sigma^\infty X & \longrightarrow & \Sigma^\infty(X/A) \\
\uparrow{\scriptstyle p^!} & & \uparrow{\scriptstyle p^!} \\
\Sigma^\infty Y & \longrightarrow & \Sigma^\infty(Y/B)
\end{array}$$

strictly commutative means that one can extend it to the following diagram.

$$\begin{array}{ccccc}
\Sigma^\infty X & \longrightarrow & \Sigma^\infty(X/A) & \longrightarrow & \Sigma^\infty(X/A) \cup C\Sigma^\infty X \\
\uparrow{\scriptstyle p^!} & & \uparrow{\scriptstyle p^!} & & \uparrow{\scriptstyle p^!} \\
\Sigma^\infty Y & \longrightarrow & \Sigma^\infty(Y/B) & \longrightarrow & \Sigma^\infty(Y/B) \cup C\Sigma^\infty Y
\end{array}$$

Now one can see what the right-hand map $p^!$ does over $C\Sigma^\infty B$ and compare it with $q^!$.

The next property concerns the composition of covering maps. Suppose given

$$X,A \xrightarrow{\ p\ } Y,B \xrightarrow{\ q\ } Z,C\ ,$$

where we assume as usual that p and q are coverings with finite fibres and $A = p^{-1}B$, $B = q^{-1}C$. Then we conclude:

$$(4.3.3) \qquad (qp)^! = p^! q^! : \Sigma^\infty(X/A) \longleftarrow \Sigma^\infty(Z/C)\ .$$

The verification is elementary.

The categorically-minded reader may be surprised that we did not state $1^! = 1$ before (4.3.3). The reason is that we have to state a stronger property, which is to "$1^! = 1$" as idempotents are to units. Suppose Y

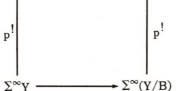

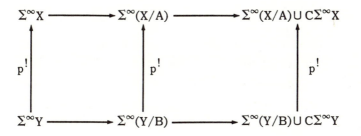

comes as a disjoint union of subcomplexes,

$$(Y,B) = \coprod_{i \in I} (Y_i, B_i) ,$$

and X comes as the disjoint union of a subset of those subcomplexes,

$$(X,A) = \coprod_{i \in J} (Y_i, B_i) .$$

The map $p : X,A \to Y,B$ is to be the injection (so that it is a fibering with 1-point fibres over Y_i for $i \in J$ and a fibering with 0-point fibres over Y_i for $i \notin J$).

(4.3.4). *The resulting map*

$$p^!: \bigvee_{i \in I} \Sigma^\infty(Y_i/B_i) \longrightarrow \bigvee_{i \in J} \Sigma^\infty(Y_i/B_i)$$

has components 1 *for* $i \in J$, 0 *for* $i \notin J$.

(This is trivial to see.)

A typical application of (4.3.4) arises when we have a covering $p : X,A \to Y,B$ in which X comes as the disjoint union of two subcomplexes X' and X'', of which X' is an n'-sheeted covering over Y and X'' is an n''-sheeted covering over Y. We apply (4.3.4) to the injections

$$X' \to X, \qquad X'' \to X$$

in conjunction with (4.3.3). We find that the map

$$\begin{array}{ccc} p^!: \Sigma^\infty(X/A) & \longleftarrow & \Sigma^\infty(Y,B) \\ \updownarrow = & & \\ \Sigma^\infty(X'/A') \vee \Sigma^\infty(X''/A'') & & \end{array}$$

has components $(p')^!$ and $(p'')^!$.

The final property we state is a product formula. Suppose given coverings

$$p': X',A' \to Y',B'$$

$$p'': X'',A'' \to Y'',B''$$

with our usual properties. Then we can form

$$p' \times p'': X' \times X'' \longrightarrow Y' \times Y''$$

and it maps $A' \times X'' \cup X' \times A''$ into $B' \times Y'' \cup Y' \times B''$. (Here one mouths the usual words about products of CW-complexes being given the CW-topology.) The quotients

$$\frac{X' \times X''}{A' \times X'' \cup X' \times A''}, \qquad \frac{Y' \times Y''}{B' \times Y'' \cup Y' \times B''}$$

may be identified with

$$(X'/A') \wedge (X''/A''), \qquad (Y'/B') \wedge (Y''/B'').$$

(4.3.5). *The map*

$$(p' \times p'')^!: \Sigma^\infty((X'/A') \wedge (X''/A'')) \longleftarrow \Sigma^\infty((Y'/B') \wedge (Y''/B''))$$

may be identified with $(p')^! \wedge (p'')^!$.

If our complexes Y', Y'' are finite-dimensional we can stick to cubes $I^{n'}, I^{n''}$ of fixed dimensions, and the verification becomes quite easy. It stays manageable if one of Y', Y'' is finite-dimensional. Beyond that it involves all the nuisances which attend smash-products of spectra; we forbear to go into that.

In order to obtain information about the behavior under transfer of internal products in cohomology, we apply (4.3.5) in conjunction with (4.3.1), using for example the following diagram.

$$\begin{array}{ccc} X,A & \xrightarrow{(1,p)} & X\times Y,\ A\times Y \\ \Big\downarrow p & & \Big\downarrow p\times 1 \\ Y,B & \xrightarrow{\Delta} & Y\times Y,\ B\times Y \end{array}$$

We find that if E is a ring-spectrum, then

$$p^!: E^*(X,A) \longrightarrow E^*(Y,B)$$

is a map of modules over $E^*(Y)$.

The reader may expect me to say something about "double coset formulae" (see e.g. [51] p. 257). I shall indeed; I advise you to avoid them. I suggest that you avoid them by thinking in geometrical terms, constructing an induced covering and seeing how it splits up into connected components. This approach can be illustrated by an example.

The example happens to involve a wreath product. I explain that $\Sigma_2 \int \Sigma_n$ will mean a subgroup of Σ_{2n}, consisting of those permutations ρ which preserve the set of pairs $(1,2), (3,4), \cdots, (2n-1,2n)$. ($\rho$ is allowed to send one pair to another and may interchange the two members of any pair.) This subgroup takes part in an exact sequence

$$1 \longrightarrow (\Sigma_2)^n \longrightarrow \Sigma_2 \int \Sigma_n \longrightarrow \Sigma_n \longrightarrow 1 .$$

EXAMPLE 4.3.6. What is the proper thing to put in the top left-hand corner of the following diagram?

$$\begin{array}{ccc} ? & \longleftarrow & H^*(B(\Sigma_2 \int \Sigma_{m+n})) \\ \Big\downarrow Tr & & \Big\downarrow Tr \\ H^*(B(\Sigma_{2m}\times\Sigma_{2n})) & \longleftarrow & H^*(B(\Sigma_{2(m+n)})) \end{array}$$

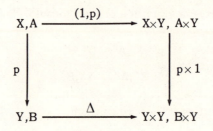

Answer. It is handy to have a model for $B\Sigma_r$ (where r will be $2(m+n)$). Consider the space of injections

$$f: \{1,2,\cdots,r\} \longrightarrow R^\infty .$$

An element of this space can be viewed as an r-tuple of distinct points in R^∞, labelled "1", "2",$\cdots$, "r"; the space is contractible, and Σ_r acts freely on it (consider composition of functions

$$R^\infty \xleftarrow{f} \{1,2,\cdots,r\} \xleftarrow{\rho} \{1,2,\cdots,r\}) .$$

Therefore this space qualifies as a space $E\Sigma_r$, and we may obtain a model for $B\Sigma_r$ by taking the appropriate quotient; we may envisage it as the space of unlabelled (and unordered) r-tuples in R^∞. We apply this with $r = 2(m+n)$. We may now form covering spaces of $B\Sigma_r$.

To form the covering space of type $B(\Sigma_{2m}\times\Sigma_{2n})$, we take the space of r-tuples in R^∞ with $2m$ of the points labelled "a" and $2n$ of the points labelled "b." The projection map

$$B(\Sigma_{2m}\times\Sigma_{2n}) \longrightarrow B(\Sigma_{2(m+n)})$$

takes a labelled r-tuple and forgets its labelling.

Similarly, to form the covering space of type $B(\Sigma_2\int\Sigma_{m+n})$, we take the space of r-tuples in R^∞ where the r-tuples come marked into $(m+n)$ pairs, though there is nothing to distinguish one pair from another, thus:

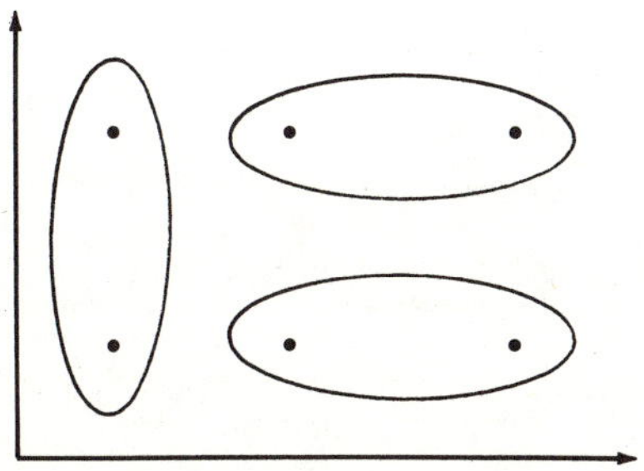
.

We now form the pullback of the diagram

$$\begin{array}{ccc} ? & \longrightarrow & B(\Sigma_2 \int \Sigma_{m+n}) \\ \big\downarrow & & \big\downarrow \\ B(\Sigma_{2m} \times \Sigma_{2n}) & \longrightarrow & B(\Sigma_{2(m+n)}) . \end{array}$$

The proper space X to go in the top left-hand corner is the space of r-tuples of distinct points in R^{∞}, where the points come marked into $(m+n)$ pairs, and also (independently) marked "a" or "b" with $2m$ a's and $2n$ b's, thus:

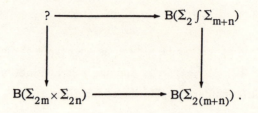

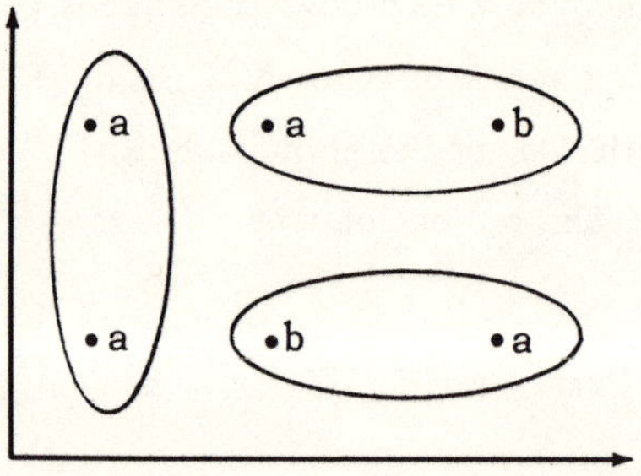

Clearly in each such marking there will be $m-q$ pairs labelled "aa," $2q$ pairs labelled "ab" and $n-q$ pairs labelled "bb." So this space X splits up as a disjoint union of components X_q corresponding to the possible values of q, $0 \leq q \leq \min(m,n)$. Here X_q is of type

$$B\big((\Sigma_2 \int \Sigma_{m-q}) \times \Sigma_q \times (\Sigma_2 \int \Sigma_{n-q})\big) .$$

So what we need in the top left-hand corner of our original diagram is

$$\bigoplus_q H^*\Big(B\big((\Sigma_2 \int \Sigma_{m-q}) \times \Sigma_q \times (\Sigma_2 \int \Sigma_{n-q})\big)\Big) ,$$

where the sum runs over $0 \leq q \leq \min(m,n)$. This answers Example 4.3.6.

Of course this still leaves the question: what do you say to the algebraist who loves double cosets and insists that this is the same thing *really*? I suggest that you smile politely and say that you are maximizing your chance of finding a helpful and congenial interpretation of the double cosets. There is no need to say that the best interpretation is one which allows you to avoid mentioning the (expletive deleted) things at all.

Finally I should point out that a version of the "double-coset formula" for Becker-Gottlieb transfer was announced by M. Feshbach at the conference in Evanston, 1977. This should be as useful when one is working with Lie groups as the classical double-coset formula is when one is working with finite groups.

CHAPTER 5

THE ADAMS CONJECTURE

§5.1. *Discussion of the conjecture*

This chapter will be divided into two sections; in the first I will explain the conjecture in question, and in the second I will discuss the proofs which have been given of it.

We have said that from the point of view of the geometry of manifolds, one of the things that topologists are required to do is to investigate the following groups and homomorphisms.

$$K_O(X) \longrightarrow K_{PL}(X) \longrightarrow K_{Top}(X) \longrightarrow K_F(X)\,.$$

We are required to give means so far as possible for computing them, to report on any phenomena which help to understand them, and so on.

The major step towards computing with a generalized cohomology theory is to compute its coefficient groups. The coefficient groups of the theory K_O are well understood; they are the homotopy groups $\pi_r(BO)$, and by the Bott periodicity theorem [42, 43, 18, 50] they are as follows.

$r =$	0	1	2	3	4	5	6	7	8 mod 8
$\pi_r(BO) =$	Z	Z_2	Z_2	0	Z	0	0	0	Z.

By contrast the coefficient groups of the theory K_F are not well understood; they are the stable homotopy groups of spheres, and while we can do calculations in some reasonable range of dimensions, we don't know a general pattern.

We can of course look at the homomorphism

$$K_O(X) \longrightarrow K_F(X)$$

which associates to each vector-bundle over X the corresponding spherical fibration. We can look at its effects on coefficient groups by putting $X = S^n$ and considering

$$\widetilde{K}_O(S^n) \longrightarrow \widetilde{K}_F(S^n) ;$$

on the left we have

$$\widetilde{K}_O(S^n) = \pi_{n-1}(O) = \pi_{n-1}(SO) \quad \text{(at least for } n > 1) ,$$

while on the right we have

$$\widetilde{K}_F(S^n) = \pi^S_{n-1}(S^0) .$$

It can be shown that the homomorphism becomes the classical J-homomorphism

$$\pi_{n-1}(SO) \longrightarrow \pi^S_{n-1}(S^0) .$$

To get any grasp of $K_F(X)$, we need invariants of spherical fibrations which are invariant under fibre-homotopy-equivalence. The first such invariants are the Stiefel-Whitney classes. The basic reason why they are fibre-homotopy invariants is that they can be defined in a particular way which I will now describe. Suppose our spherical fibrations are so arranged that over X we have E_0, the total space of the "(n–1)-sphere bundle," and it is contained in E, the total space of the "associated n-disc bundle." For example, even if $p : E_0 \to X$ is a fibration only in some weak sense, we can presumably construct E as the mapping cylinder of p; then each fibre in E is the cone of the corresponding fibre in E_0. With such arrangements we shall have a Thom isomorphism

$$\phi : H^r(X; Z_2) \longrightarrow H^{n+r}(E,E_0; Z_2) .$$

(Alternatively, E can be the total space of a "vector bundle" and E_0 the "complement of the zero section," as in §1.8. Again, if anyone prefers not to introduce E, he can replace the relative cohomology of E mod E_0 by the cohomology of the map p; but I will present matters in the conventional way.) We have the following diagram, in which Sq^i is the Steenrod square.

$$\begin{array}{ccc} H^{n+r}(E,E_0;Z_2) & \xrightarrow{Sq^i} & H^{n+r+i}(E,E_0;Z_2) \\ \uparrow{\scriptstyle \phi}\,\cong & & \uparrow{\scriptstyle \phi}\,\cong \\ H^r(X;Z_2) & & H^{r+i}(X;Z_2) \end{array}$$

We define the ith Stiefel-Whitney class of our spherical fibration ξ by

$$w_i(\xi) = \phi^{-1} Sq^i \phi(1).$$

It is more or less clear that if we replace ξ by another fibration which is fibre-homotopy-equivalent, then we replace the diagram by an isomorphic one; and so $w_i(\xi)$ does not change.

Of course the Stiefel-Whitney classes are in general not sufficient to distinguish the elements of $K_F(X)$. However, let us go back to real K-theory and consider real vector-bundles over X; then the Stiefel-Whitney classes w_1 and w_2 do enable us to select out a class of bundles which are orientable for real K-theory; if $w_1(\xi) = 0$ and $w_2(\xi) = 0$ then we have a Thom isomorphism

$$KO^r(X) \xrightarrow[\cong]{\phi} KO^{n+r}(E,E_0).$$

(At this point I concede that there is no hope of following received notation *and* being consistent; $KO(X)$ means the same as $K_O(X)$, but $KO^0(X)$ looks nicer than $K_O^0(X)$.) Anyway, this Thom isomorphism allows us to define fibre-homotopy-invariants in KO-theory, modelled on the Stiefel-Whitney classes. Of course, for this purpose we need to replace the Steenrod square Sq^i by a suitable cohomology operation on real K-theory.

In [1] I have defined certain operations

$$\Psi^k\colon KO^n(X) \longrightarrow KO^0(X)\,.$$

They are defined in terms of exterior power operations on vector-bundles; and they are so defined as to be ring-homomorphisms from the ring $KO^0(X)$ to the ring $KO^0(X)$.

We can make these operations into stable operations at the price of introducing coefficients in $Z[{}^1/_k]$; that is, we have operations

$$\Psi^k\colon KO^n(X) \longrightarrow KO^n(X; Z[{}^1/_k])$$

or

$$\Psi^k\colon KO^n(X; Z[{}^1/_k]) \longrightarrow KO^n(X; Z[{}^1/_k])\,.$$

We can therefore construct the following diagram.

$$\begin{array}{ccc} KO^n(E,E_0) & \xrightarrow{\Psi^k} & KO^n(E,E_0; Z[{}^1/_k]) \\ \phi \uparrow \cong & & \phi \uparrow \cong \\ KO^0(X) & & KO^0(X; Z[{}^1/_k])\,. \end{array}$$

We define

$$\rho^k(\xi) = \phi^{-1}\Psi^k\phi(1)\,.$$

We regard ρ^k as a characteristic class, analogous to the total Stiefel-Whitney class; it is defined on vector-bundles, and takes values in $KO^0(X; Z[{}^1/_k])$. It satisfies a sum formula, namely

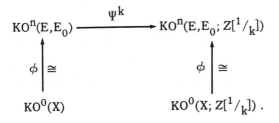

$$\rho^k(\xi \oplus \eta) = \rho^k(\xi) \cdot \rho^k(\eta) ;$$

it follows that it is defined on stable classes of vector-bundles. It has a suitable property of fibre-homotopy invariance; this is no longer quite so simple as for the Stiefel-Whitney classes, because a fibre-homotopy-equivalence need not preserve the orientation class $\phi(1)$; but one can state and prove an invariance property. We can carry out effective calculations with ρ^k. On all these points, see [3].

More generally, one should expect to see ρ^k defined on spherical fibrations with a KO-orientation – see §1.8.

This gives means for showing that two elements of $K_O(X)$ have different images in $K_F(X)$, if that should be the case. We also need means for showing that two different elements of $K_O(X)$ have the same image in $K_F(X)$, when that happens to be true. The appropriate result is as follows.

THEOREM 5.1.1. *The composite*

$$BO \xrightarrow{\psi^k - 1} BO \longrightarrow BFZ[{}^1/_k]$$

is nullhomotopic.

This statement is usually called the "Adams conjecture." It is a modern reformulation, using localization, of something labelled a conjecture in my paper [2] (see p. 183). In the statement, the map labelled $\psi^k - 1$ corresponds to the cohomology operation

$$x \mapsto \psi^k(x) - x .$$

The grounds for making such a conjecture were twofold. First, it appeared that it would enable one to set up a formal theory in which everything worked as well as it possibly could, and which allowed one to calculate the image of

$$K_O(X) \longrightarrow K_F(X)$$

by using K-theory [4]. Secondly, I was able to prove some weaker results going in the direction of (5.1.1). In particular, the following result is given in [2] (see p. 183).

PROPOSITION 5.1.2. *The composite*

$$X \xrightarrow{f} BO \xrightarrow{\psi^k - 1} BO \longrightarrow BFZ[{}^1/_k]$$

is nullhomotopic if X *is a finite complex and* f *classifies a direct sum of 2-plane bundles.*

These weaker results just sufficed to calculate the image of the classical J-homomorphism

$$\pi_{n-1}(SO) \longrightarrow \pi_{n-1}^S(S^0)$$

up to one power of 2 (or exactly if $n \not\equiv 0 \bmod 8$, but $n \equiv 0 \bmod 8$ is the interesting case). The complete result on the classical J-homomorphism follows from Theorem 5.1.1, which we are now discussing. It is perhaps interesting to note that even for the classical J-homomorphism, we have no published proof independent of the ideas I am now discussing; homotopy-theorists know in principle where they should look for an independent proof, but nobody has yet been willing to undertake the heavy task of working it out in detail and writing it down properly.

§5.2. *Proofs of the conjecture*

Theorem 5.1.1 has been proved by various authors.

A proof which is conceptually most illuminating has been given by Sullivan [150]. This proof however was not the first to appear, although it was influential long before it appeared and Sullivan deserves due credit for the influence of his ideas. Sullivan's proof uses methods from modern algebraic geometry, and I will not try to summarize it.

Sullivan writes that Quillen "first raised the specter of algebraic geometry in characteristic p in connection with the Adams conjecture." I presume this refers to the proof planned by Quillen [116] and completed by Friedlander [60]. However as Friedlander writes it the proof is for the complex case, that is, for the composite

$$BU \xrightarrow{\psi^k - 1} BU \longrightarrow BFZ[{}^1/_k] \,.$$

Of course it is always best to study the complex case first for simplicity, but you do risk losing that last power of 2 if you do not continue and obtain the results for the real case also.

The first complete proof to appear in print was Quillen's second proof [118]. I will sketch his method, considering the complex case for simplicity. It is sufficient to consider the case in which k is a prime p; but for convenience later, let it be a prime power q. Then as we have said, Quillen constructed a map

$$BGL(\infty, \bar{F}_q) \longrightarrow BU$$

which induces an isomorphism of cohomology with suitable coefficients, in particular, with coefficients in $\pi_*(BFZ[{}^1/_q])$. So by arguments from obstruction-theory, it is sufficient to prove that the composite

$$BGL(\infty, \bar{F}_q) \longrightarrow BU \xrightarrow{\psi^q - 1} BU \longrightarrow BFZ[{}^1/_q]$$

is nullhomotopic.

But now I claim that we can reduce to the following situation: we wish to prove that a composite

$$X \xrightarrow{f} BU(n) \xrightarrow{\psi^q - 1} BU \longrightarrow BFZ[{}^1/_q]$$

is nullhomotopic, where X is a finite complex, and the map f represents a bundle over X whose structural group is not U(n), but the normalizer N(T) of a maximal torus T in U(n). (This normalizer is a group extension with kernel T and with quotient the symmetric group Σ_n.) In greater detail, since the homotopy groups of $BFZ[^1/_q]$ are finite, it is sufficient to show that the composite

$$X \longrightarrow BGL(\infty, \bar{F}_q) \longrightarrow BU \xrightarrow{\Psi^q - 1} BU \longrightarrow BFZ[^1/_q]$$

is nullhomotopic, where X runs over finite subcomplexes of $BGL(\infty, \bar{F}_q)$; this one sees by standard arguments of homotopy-theory. Indeed since $GL(\infty, \bar{F}_q)$ is the union of finite subgroups $\Gamma = GL(m, F_{q^r})$, one may assume that X runs over finite subcomplexes of $B\Gamma$, where Γ runs over such finite subgroups. By a theorem of Atiyah [17, 20] any such map

$$X \to B\Gamma \to BU$$

is induced by some virtual representation α of Γ; and by induction theorems from algebra, α is a Z-linear combination of representations β induced from one-dimensional complex representations of subgroups of Γ. It is now sufficient to consider the composites

$$X \longrightarrow B\Gamma \xrightarrow{B\beta} BU(n) \xrightarrow{\Psi^q - 1} BU \longrightarrow BFZ[^1/_q] .$$

But an induced representation such as β factors in the form

$$\Gamma \to N(T) \subset U(n) .$$

This proves the claim.

Now for bundles with structural group T the result is true by (5.1.2), and it turns out that the construction which I gave in [2] to prove (5.1.2) is sufficiently natural that it works also for bundles with structural group N(T). This completes the proof I am sketching, at least for the complex case.

In his paper [118] Quillen also deals with the real case. Indeed the remarks in the last paragraph apply with little change; but the linear group $GL(\infty, \bar{F}_q)$ has to be replaced by an orthogonal group, and of course this changes the proof of cohomological equivalence.

It was the context explained above which led Quillen to his calculation of the K-groups for finite fields. In fact, the map Ψ^q of BU corresponds to a map of $BGL(\infty, \bar{F}_q)$ induced by a Frobenius automorphism $x \mapsto x^q$. It follows that the composite

$$BGL(\infty, F_q) \longrightarrow BGL(\infty, \bar{F}_q) \longrightarrow BU \xrightarrow{\Psi^q - 1} BU$$

is nullhomotopic. Let us define $F(\Psi^q - 1)$ to be the fibre of the map

$$BU \xrightarrow{\Psi^q - 1} BU$$

when this map is converted into a fibration; then we get a map

$$BGL(\infty, F_q) \longrightarrow F(\Psi^q - 1) .$$

and Quillen [119] shows that this map is a homology equivalence, so that

$$(BGL(\infty, F_q))^+ \simeq F(\Psi^q - 1) .$$

But the homotopy groups of $F(\Psi^q - 1)$ are obvious; so this calculates

$$\pi_*((BGL(\infty, F_q))^+) ,$$

that is, $K_*(F_q)$.

Finally I come to the proof of Becker and Gottlieb [33]. These authors rely on the fact that K_F is the 0th term of a generalized cohomology theory. Of course this involves quoting substantial results of infinite-loop-space theory, either from Boardman-Vogt [39, 40] or from some other author; but there is no help for it – it is essential to the argument. Pretheorem 4.1.2 now shows that the map

$$K_F(BG) \longrightarrow K_F(BN)$$

is split mono. (For the status of this argument, see Chapter 4.) Given the fact that one can make BG approximate BU or BO, and some detailed argument, this reduces the proof of the Adams conjecture to the case of a bundle with structural group $N = N(T)$; and this is handled as I have explained in discussing Quillen's proof.

This completes my survey of proofs of the Adams conjecture.

CHAPTER 6

THE SPECIAL CASE OF K-THEORY SPECTRA; THE THEOREMS OF ADAMS-PRIDDY AND MADSEN-SNAITH-TORNEHAVE

§6.1. *Introduction*

Let us begin by recalling from Chapter 1 the functor Ω^∞ from spectra to spaces. We may test the state of our knowledge by considering questions of four kinds.

(a) Given an explicit space X, is it of the form $X = \Omega^\infty \mathbf{X}$ or not? That is, is X an infinite loop space?

(b) If X is of the form $X = \Omega^\infty \mathbf{X}$, then is the spectrum $\mathbf{X}$ unique or not? That is, if X can be given the structure of an infinite loop space, is that structure unique?

(c) Given spectra $\mathbf{X}, \mathbf{Y}$ and an explicit map $f : \Omega^\infty \mathbf{X} \to \Omega^\infty \mathbf{Y}$, is f of the form $f = \Omega^\infty \mathbf{f}$ for some map $\mathbf{f} : \mathbf{X} \to \mathbf{Y}$, or not? That is, is f an infinite loop map (for the given infinite loop structures)?

(d) If in problem (c) there is a map $\mathbf{f} : \mathbf{X} \to \mathbf{Y}$ such that $\Omega^\infty \mathbf{f} = f$, is $\mathbf{f}$ unique?

As regards problem (a) we may say that the state of the art is reasonably satisfactory, and has been so for some years. There is no very significant example of a space which is suspected to be an infinite loop space but is not known to be one.

As regards (b), (c) and (d) the state of the art is not satisfactory; we possess no good method for tackling such problems in general. However, we do have interesting and useful results in particular special cases, and this will be the theme of this chapter.

Perhaps I need to amplify that remark that we don't possess general methods. Both problems (b) and (c) demand the construction of maps $f: X \to Y$. (To prove the uniqueness in problem (b), we must assume $\Omega^\infty X \simeq \Omega^\infty Y$ and construct an equivalence of spectra $e: X \to Y$.) Problem (d) is slightly worse, since it demands the construction of a homotopy. Now of course, the machine described in Chapter 2 has in principle no objection to constructing infinite loop maps. However, for this purpose you have to feed it suitable data, for example, a map between permutative categories preserving all the structure in sight. That is remarkably high octane fuel. Of course, you can have some fun in that direction, and this approach has been used in some recent work. For example, you can take the category in which the objects are the finite sets $\{1,2,\cdots,n\}$ and the maps are their automorphisms; you can then associate to each finite set the free R-module on that set as base. Or again, you can take the category in which the objects are the finite-dimensional vector spaces over the finite field F_q, and the maps are their automorphisms; you can then associate to each vector-space its underlying set. Constructions of this sort are convenient, because they lend themselves to algebraic checking and therefore lead to maps which are under good control. Topologists tend to refer to constructions of this type as "discrete models"; the reader may refer to [99] Chapter VI §5 and Chapter VIII.

However, there are many maps which one would naturally want, and which indeed exist, but which have not yet been delivered by this approach. I give an example; I am aware that some aspects of it are debatable, but historically it did play a role in the development of methods.

The functor "exterior algebra" assigns to each vector space V over R the algebra $\Lambda(V)$. It also assigns to each real vector bundle ξ over X a vector bundle $\Lambda(\xi)$; we have a canonical isomorphism

$$\Lambda(\xi \oplus \eta) \cong \Lambda(\xi) \otimes \Lambda(\eta)$$

and

$$\Lambda(1) = 2 .$$

This defines a natural transformation of K groups, or alternatively a map

$$\Lambda\colon BO_{\oplus} \longrightarrow BO_{\otimes}Z[1/2] \ .$$

One would like to know that this is an infinite loop map. Certainly, one would think, the exterior algebra is the best of all possible functors that convert addition into multiplication (see §1.8, §5.1). But the machine obstinately insists that this functor is not fit to eat; or at least, that it is not convenient to process it into a map of spectra answering the geometrical problem set.

However, it may be argued that this trouble arises from reliance on a machine which by its nature is adapted to the difficulties of the general case; perhaps we would do better to rely on the favorable circumstances of the particular cases which interest us. (Later I will say a little about what these cases may be, and why they interest people.) In practical terms, this suggestion reduces to the following: call in a consultant in old-fashioned homotopy-theory, show him the spectrum **X** and the spectrum **Y**, and request him to construct a map from **X** to **Y** by the methods of old-fashioned homotopy-theory, promising to pay him at the appropriate rates for heavy manual labor. We will see this strategy followed in §6.2, which deals with the theorem of Adams and Priddy; this theorem solves problem (b) for the particular case it studies.

However, the strategy of "appeal to homotopy-theory" sometimes runs up against a problem of communication; when you call in your consultant in old-fashioned homotopy-theory, you actually have to show him both spectra. If you merely confront him with the statement "there exists a machine-built spectrum", he is very likely to feel that this is not an animal fit for him to eat. It might be possible to present this as a reproach to machine-builders ("Make your black boxes less opaque!"); but this would not be fair. Reputable machine-builders are willing to supply all manner of invariants which reflect the infinite-loop-structure of their infinite-loop-spaces; for example, they will supply Dyer-Lashof operations and transfer in almost any theory licensed as fit for use in calculation. And no doubt it is the role of the homotopy-theorist to find out whether these "primary" invariants are sufficient to answer the problem in hand; and if they are not, it falls to him also to determine what secondary horror is needed.

Of the available invariants, it seems that so far the transfer has been the most helpful. In particular, in §6.3 we deal with the theorem of Madsen, Snaith and Tornehave, which solves problems (c) and (d) for the particular case it studies; it states that while in general transfer conditions would be necessary for the existence of a solution, in the particular case they are also sufficient. The remainder of this chapter is devoted to a careful examination of the basic ideas of the proof. The reader is warned that this is somewhat more detailed than most of the earlier parts of the book, and may call for skipping.

§6.2. *The theorem of Adams and Priddy*

This theorem answers problem (b) of §6.1 for the case which it considers. It compares an "unknown" spectrum $\mathbf{X}$ with a "standard" spectrum; to describe the latter, we introduce the method of killing homotopy groups. Let $\mathbf{Y}$ be a spectrum; then there is a spectrum which we will write $\mathbf{Y}(n,\cdots,\infty)$, which comes provided with a map $\mathbf{Y}(n,\cdots,\infty) \to \mathbf{Y}$, and which is characterized by the following two properties.

(i) $\pi_r(\mathbf{Y}(n,\cdots,\infty)) = 0$ for $r < n$.

(ii) The induced map

$$\pi_r(\mathbf{Y}(n,\cdots,\infty)) \longrightarrow \pi_r(\mathbf{Y})$$

is iso for $r \geq n$.

In particular, we take $\mathbf{Y}$ to be $\mathbf{KO}$, the spectrum representing classical, periodic, real K-theory. We then define $\mathbf{bso} = \mathbf{KO}(2,\cdots,\infty)$. Then $\mathbf{bso}$ represents "1-connected real K-theory", and $\Omega^\infty\mathbf{bso}$ is the H-space $BSO_\oplus$. We may also write our spectrum as $\mathbf{bso}_\oplus$ to distinguish this spectrum from any competitors it may have.

THEOREM 6.2.1 (Adams and Priddy, [14]). *Let* p *be a fixed prime and* X *a connective spectrum. Suppose given a homotopy equivalence of spaces*

$$\Omega^\infty X \simeq BSOZ_{(p)} .$$

Then there is a homotopy equivalence of spectra

$$X \simeq \mathbf{bso}_\oplus Z_{(p)} .$$

In this statement the space $BSOZ_{(p)}$ is the localization of the space BSO (see Chapter 3); similarly for spectra. The theorem shows that "the infinite loop space structure on $BSOZ_{(p)}$ is essentially unique".

There is a corresponding theorem in which "localization" is replaced by the more drastic step of "completion" (for which the reader may refer to [150] or [46]). There are also corresponding theorems in which SO is replaced by SU; to replace SO by O does not change the theorem provided p is odd. For the group O at the prime 2, or for the group U at any prime, the theorem would not be true as stated; but it becomes true if the hypotheses are strengthened very slightly; it is more than enough to suppose given an equivalence of H-spaces, for example

$$\Omega^\infty X \simeq BU_\oplus Z_{(p)} .$$

EXAMPLE 6.2.2. The infinite loop spaces $BSO_\oplus$ and $BSO_\otimes$ (see §1.8) become equivalent as infinite loop spaces on localization at any prime p; in other words, the corresponding spectra $\mathbf{bso}_\oplus$ and $\mathbf{bso}_\otimes$ become equivalent as spectra on localization at any prime p.

REMARK 6.2.3. The spectra $\mathbf{bso}_\oplus$ and $\mathbf{bso}_\otimes$ are not equivalent; in fact the spaces $BSO_\oplus$ and $BSO_\otimes$ are not equivalent as H-spaces. This shows that all this fuss and nonsense about localization is actually needed.

Let me sketch the proof of Remark 6.2.3. Suppose given a map of H-spaces $f: BSO_\oplus \to BSO_\otimes$. Consider the induced maps of homotopy groups

$$f_*: \pi_4(BSO) \longrightarrow \pi_4(BSO)$$
$$f_*: \pi_8(BSO) \longrightarrow \pi_8(BSO) .$$

These must be multiplication by some integers n,m respectively; and once n is fixed m is determined mod 12. We can determine the residue class of m mod 12 by looking at one well-accredited H-space map with the correct value of n; for example, with $n = 1$ we use

$$\rho^5: BSO_\oplus \longrightarrow BSO_\otimes Z[1/5]$$

(see [3]) and we find $m \equiv 7 \bmod 12$. For an equivalence we must have $m = \pm 1$, so there is no H-space equivalence with $n = 1$. The case $n = -1$ can be reduced to the case $n = +1$ by composition with the map $\Omega^\infty(-1)$ of either source or target.

EXAMPLE 6.2.4. The spaces F/PL (or F/Top) and BO become equivalent as infinite loop spaces upon localization at any odd prime p.

In fact they are infinite loop spaces, by Boardman and Vogt [39, 40] (see §1.8); and as spaces they are equivalent at any odd prime p, according to Sullivan ([150], p. 24).

Let me sketch the proof of Theorem 6.2.1.

Let X be the "unknown" spectrum, and let Y be the "standard" spectrum $bso_\oplus Z_{(p)}$, with which X has to be compared. The first major step is to prove that the cohomology $H^*(X; F_p)$ of X (with mod p coefficients) is isomorphic, as a module over the mod p Steenrod algebra A, to the expected result; for example, if $p = 2$, it has to be proved isomorphic to

$$H^*(Y; F_2) = A/ASq^3 .$$

The idea of this step is that the homotopy groups of X are determined by the known homotopy groups of $\Omega^\infty X$, and one can also get just enough information about the k-invariants of X.

The cohomology calculation allows one to look at the Adams spectral sequence

$$\mathrm{Ext}_A^{**}(H^*(Y;F_p),\,H^*(X;F_p)) \Longrightarrow [X,Y]_* \,;$$

the isomorphism of A-modules

$$H^*(Y;F_p) \cong H^*(X;F_p)$$

gives us an element of Ext^0, say θ.

The second major step is to prove that

$$\mathrm{Ext}_A^{s,t}(H^*(Y;F_p),\,H^*(X;F_p))$$

is zero for $t-s=-1$. This implies that all differentials d_r are zero on the element θ. This step is of course essentially a calculation; it is made more conceptual by developing the structure-theory of modules over small subalgebras of the Steenrod algebra A.

The third major step involves tackling the difficulties over the convergence of the Adams spectral sequence; for of course the case considered falls well outside the range of any standard set of conditions known to be sufficient for the convergence of the Adams spectral sequence. However it is possible to prove in this case that the isomorphism θ is induced by a map of spectra $f\colon X \to Y$. The proof makes essential use of the fact that we know sufficiently many endomorphisms of the standard spectrum Y. Once the map f is obtained, it is not hard to show that it is an equivalence.

This completes the sketch of the proof.

§6.3. *The theorem of Madsen, Snaith and Tornehave*

This theorem answers questions (c) and (d) of §6.1 for the case which it considers. I shall begin by stating a "global" version of the theorem, dividing it into three parts, (6.3.1)-(6.3.3). I shall then explain that there exist also "p-local" and "p-complete" versions of the theorem, and that these are more primitive and elementary, and deserve to be proved first; such results will be stated as (6.3.4)-(6.3.9), and accompanied by some discussion. Then I shall try to indicate the use of these results by mentioning some of the corollaries which Madsen, Snaith and Tornehave deduce from them; and finally I shall comment on the proofs.

Let's see some results.

THEOREM 6.3.1. *Let* X *and* Y *be connective spectra such that*

$$\Omega^\infty X \simeq BSO, \qquad \Omega^\infty Y \simeq BSO .$$

Then the function

$$\Omega^\infty : [X,Y] \longrightarrow [\Omega^\infty X, \Omega^\infty Y]$$

is injective.

Here $[X,Y]$ means homotopy classes of maps in the category of spectra, while $[\Omega^\infty X, \Omega^\infty Y]$ means homotopy classes of maps in the category of spaces. Clearly this answers problem (d) of §6.1 (for the case at issue); it says that a map of spaces $f: \Omega^\infty X \to \Omega^\infty Y$ can be given the structure of an infinite loop map in at most one way.

THEOREM 6.3.2. *Let* X *and* Y *be connective spectra such that*

$$\Omega^\infty X \simeq BSO, \qquad \Omega^\infty Y \simeq BSO .$$

Then the following conditions on a homotopy class

$$f: \Omega^\infty X \longrightarrow \Omega^\infty Y$$

are equivalent.

(i) f *lies in* $\operatorname{Im} \Omega^\infty$.

(ii) f *is a map of* H-*spaces, and the natural transformation*

$$f_*: [W,\Omega^\infty X] \longrightarrow [W,\Omega^\infty Y]$$

commutes with transfer for the covering maps

$$B(Z/_{p^r}) \longrightarrow B(Z/_{p^{r+1}})$$

(*where* p *runs over all primes and* r *runs over* $0,1,2,\cdots$).

I pause to explain the last clause. Any covering map (between connected spaces) induces a monomorphism of fundamental groups; so anyone who writes the sentence above must have in mind a monomorphism from Z_{p^r} to $Z/_{p^{r+1}}$. There is just one subgroup of $Z/_{p^{r+1}}$ isomorphic to $Z/_{p^r}$, and if we take the corresponding covering of $B(Z/_{p^{r+1}})$, we get "the" covering

$$B(Z/_{p^r}) \longrightarrow B(Z/_{p^{r+1}}) .$$

Earlier in the clause we have

$$[W,\Omega^\infty X] = [W,BSO] = \widetilde{KSO}(W) ,$$

$$[W,\Omega^\infty Y] = [W,BSO] = \widetilde{KSO}(W) ,$$

so the induced map f_* may also be written

$$f_*\colon \widetilde{KSO}(W) \longrightarrow \widetilde{KSO}(W)$$

and considered as a cohomology operation. This operation is additive if f is a map of H-spaces; the clause asks that f_* should also commute with transfer. It is supposed to be clear from Chapter 4 that condition (i) implies condition (ii).

Theorem 6.3.2 answers problem (c) of §6.1 (for the case at issue); it provides a practical test to determine whether a given map is an infinite loop map.

THEOREM 6.3.3. (a) *Let* X *and* Y *be connective spectra such that*

$$\Omega^\infty X \simeq SO, \quad \Omega^\infty Y \simeq BO ;$$

then

$$[X,Y] = 0 .$$

(b) *Let* X *and* Y *be connective spectra such that*

$$\Omega^\infty X \simeq \text{Spin}, \quad \Omega^\infty Y \simeq BSO ;$$

then

$$[X,Y] = 0 .$$

It is natural to arrive at such results if one starts from Theorems 6.3.1, 6.3.2 and asks for corresponding results about the graded group $[X,Y]_*$. Results such as Theorem 6.3.3 can be used in exact-sequence arguments to prove that other maps between spectra are unique.

There are also versions of (6.3.1), (6.3.2) in which SO is replaced by SU, and similarly for (6.3.3). For simplicity and effect I have chosen to present "global" versions of these theorems, that is, versions "over Z"; but there are also "p-local" and "p-complete" versions of the theorems, and indeed it is desirable to prove them first.

All theorems of this sort go back to Madsen, Snaith and Tornehave [86, 87]. However these authors concentrate on the p-local and the p-complete cases rather than the global case; also they do not explicitly state (6.3.3), although this follows from their methods except at the prime $p = 2$, where the result goes back to Ligaard [80]. For the global versions of (6.3.1), (6.3.2), (6.3.3) the reader may consult [99] Theorem 7.1 p. 130, Theorem 1.6 p. 212, Theorem 7.2 p. 131.

As I have said, it is advisable to prove the p-local and p-complete cases of the theorem first; and conventional wisdom says that the global version should be deduced from these by more-or-less standard use of the received machinery of localization and completion, plus perhaps $\lim^1$ arguments. I am happy to go along with this, and am accordingly suppressing the somewhat eccentric argument which appeared in the first draft of this book.

If one adopts such a proof, it is natural to make the theorem state exactly what the proof proves; and then one sees that it is not strictly necessary to assume a global equivalence

$$\Omega^\infty X \simeq BSO ;$$

it is sufficient to assume local equivalences

$$\Omega^\infty X Z_{(p)} \simeq BSOZ_{(p)}$$

(the equivalences for different primes being independent), plus the finiteness assumption that each group $\pi_r(X)$ is finitely-generated (over Z). Similarly for Y. However, the statements given above are simpler, and they are sufficient to cover the most interesting applications.

Even if we agree to forgo inspecting the standard machinery of localization as it grinds out the proof of the global case, it still seems worthwhile to go over the statement and proof of one or two of the more primitive and elementary cases to see what is involved.

Once we are dealing with a p-local or p-complete case, the theorem of Adams and Priddy, Theorem 6.2.1, can be used to replace the "unknown" spectra $\mathbf{X}$ and $\mathbf{Y}$ by "known" spectra representing versions of connective K-theory. So I will consider results stated in this way. It seems simplest to begin with the complex case, and indicate afterwards what changes are needed for the real case; moreover it seems simplest to begin with operations on standard K-theory, $K(W)$, rather than on $\widetilde{KSU}(W)$. So I reach the following prescription for the most elementary and primitive case to be studied.

Let $\mathbf{K}$ be the spectrum which represents classical, periodic, complex K-theory. By the "method of killing homotopy groups" (see §6.2) we construct $\mathbf{K}(0,\cdots,\infty)$; this spectrum represents connective complex K-theory. It is often written **bu**, but I will write it **ku** (for reasons given in [99] p. 121). I take

$$\mathbf{X} = \mathbf{ku} = \mathbf{K}(0,\cdots,\infty) .$$

Then

$$\Omega^\infty X = Z \times BU ,$$

$$[W, \Omega^\infty X] = K(W) .$$

Let Λ be an abelian group of coefficients which is torsion-free. Our main interest lies with $\Lambda = Z_{(p)}$, the ring of integers localized at p, and $\Lambda = Z_p^\wedge$, the ring of p-adic integers; but there is no reason why one should not consider other groups. By introducing coefficients (see Chapter 3) we can form

$$Y = X\Lambda = \mathbf{ku}\Lambda .$$

For example, if $\Lambda = Z_{(p)}$, then $Y = X\Lambda = \mathbf{ku}\Lambda$ is the localization of $X = \mathbf{ku}$ at p. The use of coefficients in $Z_p^\wedge$ serves as a substitute for "completion" in other treatments. As an alternative to forming $\mathbf{ku}\Lambda$, we may first afflict $\mathbf{K}$ with coefficients so as to get $\mathbf{K}\Lambda$, and afterwards kill homotopy groups so as to get

$$(\mathbf{K}\Lambda)(0,\cdots,\infty) ;$$

this comes to the same thing and gives a spectrum equivalent to $\mathbf{ku}\Lambda$. We may write

$$\Omega^\infty Y = \Lambda \times BU\Lambda$$

as a definition of the space $BU\Lambda$. We have

$$[W, \Omega^\infty Y] = K\Lambda(W) ,$$

where the right-hand side is "K-theory with coefficients in Λ."

PROPOSITION 6.3.4. *Theorem 6.3.1 remains true for this choice of* X *and* Y; *that is,*

$$\Omega^\infty \colon [X,Y] \longrightarrow [\Omega^\infty X, \Omega^\infty Y]$$

is injective.

PROPOSITION 6.3.5. *Theorem 6.3.2 remains true for this choice of* X *and* Y, *provided that* $\Lambda = Z_{(p)}$ *or* $\Lambda = Z_p^\wedge$.

In fact it is enough to consider coverings

$$B(Z/_{p^r}) \longrightarrow B(Z/_{p^{r+1}})$$

for the same prime p that appears in $\Lambda = Z_{(p)}$ or $\Lambda = Z_p^\wedge$; of course r must still run over $0,1,2,\cdots$. If (6.3.5) uses fewer coverings than (6.3.2), then it is reasonable to prove (6.3.5) before (6.3.2).

Now we want to pass on to cases in which $\Omega^\infty X$ is BSU rather than $Z \times BU$.

PROPOSITION 6.3.6. *Theorem 6.3.1 remains true if we take*

$$X = \mathbf{K}(2n,\cdots,\infty) \text{ for any } n \geq 0, \text{ and}$$
$$Y = X\Lambda \text{ where } \Lambda \text{ is torsion-free.}$$

Clearly this includes (6.3.4), and gives a result such as we seek by taking $n = 2$. The spectrum $\mathbf{K}(4,\cdots,\infty)$ represents 3-connected complex K-theory, and may be written **bsu** because

$$\Omega^\infty X = BSU .$$

PROPOSITION 6.3.7. *Theorem 6.3.2 remains true if we take*

$$X = \mathbf{K}(4,\cdots,\infty) ,$$
$$Y = X\Lambda \text{ where } \Lambda = Z_{(p)} \text{ or } Z_p^\wedge .$$

Again it is enough to consider coverings

$$B(Z/_{p^r}) \longrightarrow B(Z/_{p^{r+1}})$$

for the prime p in question.

Next we want to pass on to real K-theory. Let us replace $\mathbf{K}$ by the spectrum **KO** which represents classical, periodic, real K-theory. By the method of killing homotopy groups we construct

$$X = KO(2,\cdots,\infty) ;$$

this spectrum represents 1-connected real K-theory, and may be written bso because

$$\Omega^\infty X = BSO .$$

PROPOSITION 6.3.8. *Theorem 6.3.1 remains true for* $X = \mathbf{bso}$ *and* $Y = \mathbf{bso}\Lambda$, *where* Λ *is torsion-free.*

PROPOSITION 6.3.9. *Theorem 6.3.2 remains true for* $X = \mathbf{bso}$ *and* $Y = \mathbf{bso}\Lambda$, *where* $\Lambda = Z_{(p)}$ *or* $Z_p^\wedge$.

Again (of course) it is sufficient to consider coverings

$$B(Z/_{p^r}) \longrightarrow B(Z/_{p^{r+1}})$$

for the prime p in question.

I will now try to indicate the use of these results by giving some of the corollaries which Madsen, Snaith and Tornehave deduce. I will not sketch the proofs here.

COROLLARY 6.3.10. *The map*

$$\psi^k\colon BSO_{\otimes(p)} \longrightarrow BSO_{\otimes(p)}$$

is an infinite loop map if $k \not\equiv 0 \bmod p$.

Here $X_{(p)}$ means the same as $XZ_{(p)}$ in Chapter 3 and earlier in this section. For this result, see [86], Theorem 4.5, p. 39. However, May has pointed out that this result can be deduced from the theorem of Adams and Priddy without using the main result of Madsen, Snaith and Tornehave; see [99], Lemma 7.6, p. 132.

COROLLARY 6.3.11. *The map*

$$\rho^k: BSO_{\oplus(p)} \longrightarrow BSO_{\otimes(p)}$$

is an infinite loop map if $k \not\equiv 0 \bmod p$.

Here it should be pointed out that the symbol ρ^k can mean different things; see [86] pp. 36-37. The meaning which comes naturally to me is that explained in Chapter 5, and indeed this ρ^k can be defended as useful. With this interpretation, the reference is [86] Corollary 4.4 p. 38. Other results of [86], in particular Proposition D, p. 4 and Theorem 4.3, p. 37 are forms of (6.3.11) with a different interpretation of ρ^k.

COROLLARY 6.3.12. *The maps*

$$e: F/O \longrightarrow BSO_{\otimes}$$

$$\sigma: F/PL \xrightarrow{\simeq} BSO_{\otimes}Z[1/2]$$

defined by Sullivan [145] *are both infinite loop maps.*

See [86] Theorem E, p. 5.

I now turn to the question of proofs. The object of the next three sections will be to examine or explain proofs of Propositions 6.3.4 - 6.3.9. It will be simplest to concentrate on the proofs of Propositions 6.3.4 and 6.3.5, and point out afterwards what changes or extra arguments are needed to prove Propositions 6.3.6 - 6.3.9.

The proof which I offer for Proposition 6.3.4 will be given in §6.4; it is different from the proof of Madsen, Snaith and Tornehave, and relies on a universal coefficient theorem; it can be made to prove Theorem 6.3.3 also.

I turn to sketch the proof of Proposition 6.3.5. Let $A(\Lambda)$ be the set of H-space maps

$$a: Z \times BU \longrightarrow \Lambda \times BU\Lambda$$

considered as additive cohomology operations

$$a: \mathbf{K}(W) \longrightarrow \mathbf{K}\Lambda(W) .$$

Then the structure of $A(\Lambda)$ can be determined quite easily; this is done in §6.4 as Lemma 6.4.1, because the result comes in handy in that section. So much works whenever Λ is torsion-free; but for the rest of the proof, it is best to begin with the p-adic case $\Lambda = Z_p^\wedge$. (The deduction of the p-local case $\Lambda = Z_{(p)}$ from the p-adic case $\Lambda = Z_p^\wedge$ is done at the end of §6.4.) In fact I shall assume that the coefficient group Λ is complete and Hausdorff for its p-adic topology. It is then particularly easy to describe the structure of $\mathbf{K}\Lambda(W)$, where $W = B(Z/_{p^r})$ or $B(Z/_{p^{r+1}})$; see Lemma 6.5.1. We can then find the condition for an element $a \in A(\Lambda)$ to commute with the transfer for the covering maps

$$\begin{array}{ccc} B(1) & \longrightarrow & B(Z/_p) \\ B(Z/_p) & \longrightarrow & B(Z/_{p^2}) \\ & \vdots & \\ B(Z/_{p^{r-1}}) & \longrightarrow & B(Z/_{p^r}) . \end{array}$$

This is the essential business of §6.5, and it is not difficult either; the condition is that $a \in A(\Lambda)$ has the same effect in $B(Z/_{p^r})$ as some finite Λ-linear combination

$$a_r = \sum_{k \not\equiv 0 \bmod p} \lambda_k \psi^k$$

of operations ψ^k with k prime to p. (See Lemma 6.5.6.)

Now it is well known (as we have already mentioned in §5.1) that $\lambda_k \psi^k$ is an infinite loop map provided we use a coefficient group Λ such that $k: \Lambda \to \Lambda$ is iso; and in our case this holds when k is prime to p.

It follows that any finite sum

$$a_r = \sum_{k \not\equiv 0 \bmod p} \lambda_k \Psi^k$$

is also an infinite loop map.

So we reach the conclusion that our given element $a \epsilon A(\Lambda)$ can be approximated (in a suitable sense) by elements a_r, and that each a_r can be lifted to a map of spectra

$$b_r \epsilon [X,Y] .$$

What we need, however, is a single element

$$b \epsilon [X,Y]$$

such that $\Omega^\infty b = a$. At this point [86], p. 20, lines 4-5 and [87], p. 410, lines 6-7 mention "a slight convergence problem, which we safely leave to the reader." It is always flattering to the reader to know that the authors have such confidence in him. However, the sceptical reader may perhaps wonder whether this is not the point in the argument which most requires proof, and whether he can "safely" leave such a "slight convergence problem" to the authors. In §6.6 I will therefore spend the time to examine this convergence problem. Needless to say I reach the thankless conclusion that the authors of [86] speak the precise truth; they can safely leave this problem to me. I suspect however that the argument I shall give is quite different from the one the authors of [86] have in mind, and reflects a difference of taste. More precisely, I imagine that the authors of [86] have it in mind to rely on the properties of "completion in the sense of Sullivan," or some equivalent construction. Now completion in the sense of Sullivan is essentially compactification; if it is applicable at all it leads to topologies which are compact. Moreover topologies which are compact are exceedingly convenient, and I suspect that the authors of [86] have it in mind to exploit that fact. But looking at the mathematics, it seems to me that compactness is neither necessary nor relevant; to give a silly example, the results are true for

$$\Lambda = \bigoplus_{1}^{\infty} Z_p^{\wedge}$$

although this group is not compact. However, completeness is highly relevant. Since I am interested in working out what really goes on, I shall write in terms of completeness rather than compactness, even if it takes longer.

§6.4. *Calculation of* $[\mathbf{ku}, \mathbf{ku}\Lambda]$ *and proof of Proposition 6.3.4*

Throughout this section I shall assume that the group of coefficients Λ is torsion-free, but not necessarily complete or Hausdorff. The object of this section is to do all the work which depends only on the torsion-free assumption, before any other assumption is introduced. To begin with I shall give the structure of the group $A(\Lambda)$ introduced in §6.3. After that I give means for computing $[\mathbf{ku}, \mathbf{ku}\Lambda]$; this leads to a proof of Proposition 6.3.4. Finally I show that in Proposition 6.3.5, the case $\Lambda = Z_{(p)}$ follows from the case $\Lambda = Z_p^{\wedge}$. At intervals I shall comment on the extension of these methods to the other cases mentioned in (6.3.6)-(6.3.9).

The group $A(\Lambda)$ was defined to be the set of H-space maps

$$a\colon Z \times BU \longrightarrow \Lambda \times BU\Lambda ,$$

or equivalently the set of additive cohomology operations

$$a\colon \mathbf{K}(W) \longrightarrow \mathbf{K}\Lambda(W) .$$

In order to study $A(\Lambda)$, we use CP^{∞} as a test space. Let $\xi \in \mathbf{K}(CP^{\infty})$ be the element represented by the universal line bundle over CP^{∞}, that is by the obvious map

$$CP^{\infty} = BU(1) \longrightarrow BU = 1 \times BU \subset Z \times BU .$$

Let $x = \xi - 1$. Then $K\Lambda(CP^\infty)$ may be identified with the additive group $\Lambda[[x]]$ of formal power-series $\sum_{i \geq 0} \lambda_i x^i$. Thus an element $a \in A(\Lambda)$ gives us a formal power-series

$$a(\xi) = \sum_{i \geq 0} \lambda_i x^i .$$

LEMMA 6.4.1. *This construction gives an isomorphism from* $A(\Lambda)$ *to* $K\Lambda(CP^\infty) \cong \Lambda[[x]]$.

This is a standard result. See for example [6] pp. 85-87; the hypotheses on Λ assumed there are unnecessarily restrictive, but it hardly affects the argument.

COROLLARY 6.4.2. *An element* $a \in A(\Lambda)$ *is determined by the induced maps of homotopy groups*

$$a_*: \pi_{2j}(Z \times BU) \longrightarrow \pi_{2j}(\Lambda \times BU\Lambda) \qquad (j \geq 0) .$$

Proof. For each i we can construct a fixed element $b_i \in A(Z)$ such that

$$b_i(\xi) = x^i .$$

(This follows from Lemma 6.4.1, but the construction is explicitly given when one proves that lemma; see [6] p. 86.) The induced map

$$(b_i)_*: \pi_{2j}(Z \times BU) \longrightarrow \pi_{2j}(Z \times BU)$$

is multiplication by an integer β_{ij}, which is zero if $j < i$, and non-zero if $j = i$; in fact an easy calculation with the Chern character shows that

$$\beta_{ii} = i! .$$

Take then an element $a \in A(\Lambda)$ such that

$$a(\xi) = \sum_i \lambda_i x^i ;$$

the induced map

$$a_*: \pi_{2j}(Z \times BU) \longrightarrow \pi_{2j}(\Lambda \times BU\Lambda)$$

is multiplication by

$$\mu_j = \sum_{i=0}^{j} \lambda_i \beta_{ij} .$$

Since the matrix $[\beta_{ij}]$ is non-singular, the μ's determine the λ's. This proves the lemma.

We shall need the details of this proof again in §6.6.

The preceding results carry over to the real case. The structure of $\mathbf{KO}\Lambda^0(CP^\infty)$ is known [15, 123]. To obtain convenient "generators," one defines $\eta^{(i)}$ to be the real 2-plane bundle underlying the complex bundle ξ^i, or alternatively one takes the "realification" of the powers x^i. We can define $AO(\Lambda)$ to be the set of H-space maps

$$a: Z \times BSO \longrightarrow \Lambda \times BSO\Lambda ;$$

we get an isomorphism

$$AO(\Lambda) \longrightarrow 2\Lambda \oplus \widetilde{\mathbf{KO}\Lambda}^0(CP^\infty)$$

by

$$a \longmapsto a(\eta) .$$

(The factor 2 in 2Λ arises because η is of augmentation 2.) Corollary 6.4.2 remains true for the real case.

I have already said that to compute $[\mathbf{ku}, \mathbf{ku}\Lambda]$ I propose to rely on a universal coefficient theorem. For this purpose we must begin by obtaining results on the K-homology groups $K_*(\mathbf{ku})$.

LEMMA 6.4.3. $K_*(ku)$, *considered as a left module over* $\pi_*(K)$, *is free on (countably many) generators of degree* 0.

This will follow from the next lemma plus one reference.

LEMMA 6.4.4. *The map*

$$K_*(ku) \longrightarrow K_*(K)$$

(induced by $ku \to K$*) is mono.*

Proof of Lemma 6.4.3 from Lemma 6.4.4. $K_*(K)$, considered as a left module over $\pi_*(K)$, is free on countably many generators of degree 0 by a theorem of Adams and Clarke [11]. But $\pi_*(K)$ is a (graded) principal ideal domain, so any submodule of a free module over $\pi_*(K)$ is free.

As an alternative to this argument, one may of course apply the method of [11] directly to the case of $K_*(ku)$. As one still has to prove Lemma 6.4.4, there is no saving.

A third option (probably less convenient) is to base the calculation on the method of [9] pp. 331-371.

Lemma 6.4.3 remains true for the real case: $KO_*(bso)$ is a free module over $\pi_*(KO)$ on (countably many) generators of degree zero. Of course we lose the argument that "a submodule of a free module is free" because $\pi_*(KO)$ is not a principal ideal domain; we have to apply the method of [11] directly to the case of $KO_*(bso)$. The essential ideas are those given below; they require a few extra arguments to deal with the 2-torsion in $\pi_i(KO)$ for $i \equiv 1,2 \bmod 8$, and these will be omitted.

It is perhaps worth recording that this method is economical on data. The result that $KO_*(X)$ is free remains true if X is as in Theorem 6.3.1, or even if we are only given local equivalences

$$\Omega^\infty X Z_{(p)} \simeq BSOZ_{(p)}$$

plus the finiteness assumption that each group $\pi_r(X)$ is finitely-generated over Z. So what we are doing is not unduly restricted in its scope.

It seems best to amplify Lemma 6.4.4 before we prove it. To this end I recall the description of $K_*(K)$ given in [12]. This is based on the embedding

$$K_*(K) \longrightarrow K_*(K) \otimes Q ,$$

so I have to describe the group $K_*(K) \otimes Q$. The ring $\pi_*(K)$ is the ring of finite Laurent series $Z[t, t^{-1}]$, where t is the generator in $\pi_2(K)$. It follows that $K_*(K) \otimes Q$ is the ring of finite Laurent series

$$Q[u,v,u^{-1},v^{-1}] ,$$

where u and v come from t by using the maps

$$\pi_*(K) = K_*(\Sigma^\infty S^0) \longrightarrow K_*(K)$$

$$\pi_*(K) = (\Sigma^\infty S^0)_* K \longrightarrow K_*(K) .$$

Then $K_*(K)$ is described in [12] as the subgroup of finite Q-linear combinations

$$\sum_{r,s} \lambda_{r,s} u^r v^s \in Q[u,v,u^{-1},v^{-1}]$$

which satisfy certain integrality conditions.

Even to prove Lemma 6.4.4, it is useful to consider the spectra $K(2n,\cdots,\infty)$ between K and $\mathbf{ku} = K(0,\cdots,\infty)$. (For this notation, see §6.2.)

LEMMA 6.4.5. *The induced map*

$$K_*(K(2n,\cdots,\infty)) \longrightarrow K_*(K)$$

is mono, and its image is the set of sums $\sum_{r,s} \lambda_{r,s} u^r v^s$ *in* $K_*(K)$ *such that* $\lambda_{r,s} = 0$ *for* $s < n$.

This clearly implies Lemma 6.4.4.

Proof of Lemma 6.4.5. It is clear that the image of $K_*(K(2n,\cdots,\infty))$ is contained in the set of sums described, because $K_*(K(2n,\cdots,\infty))\otimes Q$ is the set of sums $\sum_{r,s} \lambda_{r,s} u^r v^s$ in which s runs over the range $s \geq n$.

We now consider the cofibering

$$K(2n+2,\cdots,\infty) \xrightarrow{\ i\ } K(2n,\cdots,\infty) \xrightarrow{\ j\ } EM(Z,2n)\ ,$$

where $EM(Z,2n)$ means an Eilenberg-MacLane spectrum of type $(Z,2n)$. By periodicity it is sufficient to consider it for $n = 0$, and we obtain the following commutative diagram.

$$\begin{array}{ccccc}
K_*(K(2,\cdots,\infty)) & \xrightarrow{\ i_*\ } & K_*(K(0,\cdots,\infty)) & \xrightarrow{\ j_*\ } & K_*(EM(Z,O)) \\
 & & \downarrow & & \downarrow{\scriptstyle \alpha} \\
 & & K_*(K) & \xrightarrow{\ (ch_0)_*\ } & K_*(EM(Q,O)) \\
 & & & & \updownarrow{\scriptstyle \cong} \\
 & & & & \pi_*(K)\otimes Q\ .
\end{array} \tag{6.4.6}$$

Here the map

$$ch_0 : K \longrightarrow EM(Q,O)$$

is the 0th component of the Chern character; the composite

$$K_*(K) \longrightarrow \pi_*(K)\otimes Q$$

is described by

$$u^r v^s \mapsto 0 \ \text{ if } \ s \neq 0$$
$$u^r v^s \mapsto t^r \ \text{ if } \ s = 0\ .$$

The map α is induced by the obvious map

$$EM(Z,O) \longrightarrow EM(Q,O) ,$$

and I begin by showing that α is iso. In fact, by the observation of G. W. Whitehead, we have

$$K_*(EM(Z_p,O) \cong H_*(K; Z_p) ,$$

and it is well known that

$$H_*(K; Z_p) = 0$$

for any prime p. Passing to extensions, we deduce that

$$K_*(EM(G,O)) = H_*(K; G) = 0$$

for any finite abelian group G. Passing to direct limits, the same conclusion holds when G is a torsion group, in particular when $G = Q/Z$. The cofibering

$$EM(Z,O) \longrightarrow EM(Q,O) \longrightarrow EM(Q/Z,O)$$

gives an exact sequence

$$\cdots \longrightarrow K_*(EM(Z,O)) \xrightarrow{\alpha} K_*(EM(Q,O)) \longrightarrow K_*(EM(Q/Z,O)) = 0 ,$$

and we see that α is iso, as claimed.

I will now use this to prove that the map j_* in Diagram 6.4.6 is epi.

By using the injection $BU(1) \to BU$ and by regarding the space BU as term 2 in the spectrum **ku**, we can construct an element in $K_0(\mathbf{ku})$ whose image in $K_0(\mathbf{K})$ is

$$\frac{1}{n!}(u^{-1}v-1)(u^{-1}v-2)\cdots(u^{-1}v-(n-1)) .$$

This calculation was first given in [12], and has been simplified in other work since (e.g. [13]). The slight difference from the notation of [12] p. 407 is due to two causes. First, in the work above I have used the space BU as term 2 in the spectrum instead of term 0 as in [12]. Secondly I have used $K_0(\ )$ instead of $K_{2n}(\ )$. At all events, the image of our element under the map

$$K_0(K(0,\cdots,\infty)) \longrightarrow \pi_0(K)\otimes Q$$

is

$$(-1)^{n-1}/_n \, .$$

Since this holds for each n, it proves that the map j_* in Diagram (6.4.6) is epi.

The exact sequence

$$K_*(K(2,\cdots,\infty)) \xrightarrow{i_*} K_*(K(0,\cdots,\infty)) \xrightarrow{j_*} K_*(EM(Z,O))$$

now shows that the map i_* is mono. By periodicity, it follows that

$$K_*(K(2n+2,\cdots,\infty)) \longrightarrow K_*(K(2n,\cdots,\infty))$$

is mono for each n.

We now consider the spectrum K as the limit of the sequence of spectra

$$\cdots \longrightarrow K(2n+2,\cdots,\infty) \longrightarrow K(2n,\cdots,\infty) \longrightarrow \cdots .$$

Thus $K_*(K)$ is the direct limit of the direct system

$$\cdots \longrightarrow K_*(K(2n+2,\cdots,\infty)) \longrightarrow K_*(K(2n,\cdots,\infty)) \longrightarrow \cdots .$$

Since this is a direct system of monomorphisms, it follows that the map from each term to the limit is mono. This proves the first assertion in Lemma 6.4.5, which incidentally proves Lemmas 6.4.4 and 6.4.3.

To complete the proof, consider an element in $K_*(K)$ of the form

$\sum_{r,s} \lambda_{r,s} u^r v^s$, where $\lambda_{r,s} = 0$ for $s < n$. Since $K_*(K)$ is the direct limit of the system considered above, this element must come from some term, say from

$$x \in K_*(K(2m, \cdots, \infty)) .$$

Now we argue by induction over m. If $m < n$ we have the following commutative diagram, analogous to Diagram (6.4.6).

$$\begin{array}{ccccc}
K_*(K(2m+2,\cdots,\infty)) & \xrightarrow{i_*} & K_*(K(2m,\cdots,\infty)) & \xrightarrow{j_*} & K_*(EM(Z,2m)) \\
 & & \downarrow & & \downarrow a \\
 & & K_*(K) & \xrightarrow{(ch_m)_*} & K_*(EM(Q,2m)) \\
 & & & & \updownarrow \cong \\
 & & & & \pi_*(K) \otimes Q
\end{array}$$

Since the image of x in $K_*(K)$ is $\sum_{r,s} \lambda_{r,s} u^r v^s$ with $\lambda_{r,s} = 0$ for $s = m$, we see that x maps to 0 in $\pi_*(K) \otimes Q$; thus $x \in \operatorname{Ker} j_*$ and $x \in \operatorname{Im} i_*$. This completes the induction, and shows that our element comes from $K_*(K(2n, \cdots, \infty))$. This proves Lemma 6.4.5.

The argument of this proof (which is purely stable and involves groups easy to describe) serves as a substitute for the use which Madsen, Snaith and Tornehave make of results on the K-cohomology of Eilenberg-MacLane spaces.

LEMMA 6.4.7. *There is an isomorphism*

$$[\mathbf{ku}, K\Lambda]_* \longrightarrow \operatorname{Hom}^*_{\pi_*(K)} (K_*(\mathbf{ku}), \pi_*(K\Lambda)) ;$$

it sends $f \in [\mathbf{ku}, K\Lambda]_* = K\Lambda^*(\mathbf{ku})$ *to the homomorphism given on* $x \in K_*(\mathbf{ku})$ *by* $x \mapsto \langle f, x \rangle$.

Here $\langle f,x \rangle$ is the Kronecker product of the element f in $K\Lambda$-cohomology and the element x in K-homology.

Proof. This is the form which the universal coefficient theorem takes in our case. More precisely, the universal coefficient theorem gives a spectral sequence

$$\mathrm{Ext}^{**}_{\pi_*(K)}(K_*(W),\ \pi_*(K\Lambda)) \Longrightarrow K\Lambda^*(W)\,,$$

in which the edge-homomorphism is given by the formula above; it sends $f \in K\Lambda^*(W)$ to the homomorphism $K_*(W) \to \pi_*(K\Lambda)$ given on $x \in K_*(W)$ by $x \mapsto \langle f,x \rangle$. For this form of the universal coefficient theorem, see [6] Lecture 1; this source contains nothing about the convergence of the spectral sequence, but that may be proved following the indications given in [7] p. 11. The spectral sequence is convergent in the sense that it satisfies Theorem 8.2 of [9] p. 224. In our case, the groups $\mathrm{Ext}^{s,*}_{\pi_*(K)}(K_*(ku), \pi_*(K\Lambda))$ are zero for $s > 0$ by Lemma 6.4.3, and so the spectral sequence reduces to its edge-homomorphism. This proves Lemma 6.4.7.

Lemma 6.4.7 remains true for the real case; there is an isomorphism

$$[bso, KO\Lambda]_* \longrightarrow \mathrm{Hom}^*_{\pi_*(KO)}(KO_*(bso),\ \pi_*(KO\Lambda))\,.$$

The proof is the same.

Another option (probably less convenient) for computing the groups $[ku, K\Lambda]_*$ and $[bso, KO\Lambda]_*$ is to follow the method of [14].

COROLLARY 6.4.8. (a) $[ku, K\Lambda]_*$ *is zero in odd degrees.*

(b) *If* $f \in [ku, K\Lambda]$, *and if*

$$f_*: \pi_{2s}(ku) \longrightarrow \pi_{2s}(K\Lambda)$$

is zero for all $s \geq 0$, *then* $f = 0$.

Proof. Part (a) follows immediately from Lemma 6.4.7, since the group

$$\mathrm{Hom}^*_{\pi_*(K)}\ (K_*(ku),\ \pi_*(K\Lambda))$$

is zero in odd degrees by Lemma 6.4.3.

For part (b), suppose given a map $f \in [ku, K\Lambda]$ such that every composite

$$\Sigma^\infty S^{2s} \xrightarrow{t^s} ku \xrightarrow{f} K\Lambda$$

(with $s \geq 0$) is zero. With the notation introduced for Lemma 6.4.5, this shows that

$$<f, u^r v^s> = 0 \text{ for } s \geq 0 .$$

If $x \in K_*(ku)$, then by Lemma 6.4.5 we can write some non-zero integer multiple νx of x in the form

$$\nu x = \sum_{s \geq 0} \lambda_{r,s} u^r v^s$$

where $\lambda_{r,s} \in Z$. Then $<f, \nu x> = 0$, and we deduce $<f,x> = 0$ because Λ is torsion-free. Now the conclusion $f = 0$ follows by Lemma 6.4.7. This proves Corollary 6.4.8.

Corollary 6.4.8(a) carries over to the real case in the sense that $[bso, KO\Lambda]_*$ is zero in degree -1. The proof is the same; $KO_*(bso)$ has its generators in degree 0 and $\pi_*(KO\Lambda)$ is zero in degree -1, so the Hom^* group is zero in degree -1. This provides the means to prove theorems of the general type of (6.3.3a); similarly for (6.3.3b).

Corollary 6.4.8(b) also carries over to the real case, with the trivial change that π_{2s} becomes π_{4s}. In particular, a map f: $bso \to KO\Lambda$ is determined by its induced map of homotopy groups. The proof is the same.

Proof of Proposition 6.3.4. By standard connectivity arguments we have

$$[ku, ku\Lambda] \xrightarrow{\cong} [ku, K\Lambda] ;$$

moreover, the effect of $f: ku \to ku\Lambda$ on homotopy groups can be read off from the effect of $\Omega^\infty f: Z \times BU \to \Lambda \times BU\Lambda$ on homotopy groups. Thus (6.3.4) follows from (6.4.8b).

Proof of Proposition 6.3.6. Lemma 6.4.3 remains true if **ku** is replaced by $K(2n,\cdots,\infty)$ for any n (either because this follows from (6.4.3) by periodicity, or because the proof of (6.4.3) works equally well in the general case, given (6.4.5)). Therefore Lemmas 6.4.7 and 6.4.8 remain true if **ku** is replaced by $K(2n,\cdots,\infty)$; the proofs are hardly affected. Now the proof of (6.3.4) carries over to prove (6.3.6).

Similarly, these arguments carry over to the real case to prove Proposition 6.3.8.

LEMMA 6.4.9. *If Proposition 6.3.5 is true for the p-adic case* $\Lambda = Z_p^\wedge$, *then it is true for the p-local case* $\Lambda = Z_{(p)}$.

Proof. Let $a \in A(Z_{(p)})$. Then the induced map of homotopy groups

$$a_*: \pi_{2s}(Z \times BU) \longrightarrow \pi_{2s}(Z_{(p)} \times BUZ_{(p)})$$

is multiplication by some scalar $\lambda_s \in Z_{(p)}$. We have to assume that a commutes with transfer and that Proposition 6.3.5 is true for $\Lambda = Z_p^\wedge$; then there is a map of spectra

$$f: ku \longrightarrow kuZ_p^\wedge$$

such that $\Omega^\infty f$ is the obvious composite

$$Z \times BU \xrightarrow{a} Z_{(p)} \times BUZ_{(p)} \longrightarrow Z_p^\wedge \times BUZ_p^\wedge .$$

Then the induced map of homotopy groups

$$f_*: \pi_{2s}(ku) \longrightarrow \pi_{2s}(kuZ_p^\wedge)$$

is multiplication by the same scalar $\lambda_s \in Z_{(p)}$. It follows that for each element $u^r v^s$ we have

$$<f, u^r v^s> \in \pi_*(K) \otimes Z_{(p)} .$$

Let $x \in K_*(ku)$; using Lemma 6.4.5 as in the proof of (6.4.8b), we can write

$$\nu x = \sum \lambda_{r,s} u^r v^s \quad \text{with} \quad \lambda_{r,s} \in Z ,$$

and hence

$$<f, \nu x> \in \pi_*(K) \otimes Z_{(p)} .$$

But now it follows that the element $<f,x>$, which *a priori* lies in $\pi_*(K) \otimes Z_p^\wedge$, actually lies in $\pi_*(K) \otimes Z_{(p)}$. So Lemma 6.4.7 shows that f factors through a map

$$f': ku \to kuZ_{(p)} .$$

Finally we deduce $\Omega^\infty f' = a$ by Corollary 6.4.2. This proves Lemma 6.4.9.

The same argument serves to deduce the case $\Lambda = Z_{(p)}$ of Proposition 6.3.7 from the case $\Lambda = Z_p^\wedge$, but we have to pay for the fact that we have not said anything much about the analogue of $A(\Lambda)$ for $\widetilde{KSU}$-theory. In fact, at the end of the argument, we need to know that the given map

$$a: BSU \longrightarrow BSUZ_{(p)}$$

is determined by the composite

$$BSU \xrightarrow{a} BSUZ_{(p)} \longrightarrow BSUZ_p^\wedge .$$

However, we can easily prove this – for example, from the Atiyah-Hirzebruch spectral sequence. I don't recommend trying that argument on BSO to deduce the case $\Lambda = Z_{(p)}$ of Proposition 6.3.9 from the case

$\Lambda = Z_p^\wedge$, but then, I did say just enough about the analogue of $A(\Lambda)$ for the real case.

It is sufficiently clear that the argument of (6.4.9) is not restricted to the particular groups $Z_{(p)}$ and $Z_p^\wedge$; under appropriate assumptions on Λ we can certainly work with the embedding of Λ in its p-adic completion.

§6.5. *The transfer calculation*

In this section and the next I shall assume that the group of coefficients Λ is torsion-free, and that it is complete and Hausdorff for its p-adic topology; for example, $\Lambda = Z_p^\wedge$ will do. In this section I shall compute the groups $K\Lambda(B(Z/p^r))$, and work out the effect of the assumption that an element $a \in A(\Lambda)$ commutes with transfer. The result is expressed in terms of a direct-sum splitting of $A(\Lambda)$.

I begin with the calculation of $K\Lambda(B(Z/p^r))$. There is a unique subgroup of $U(1)$ isomorphic to Z/p^r; the embedding defines a map

$$B(Z/p^r) \longrightarrow BU(1) = CP^\infty .$$

Let $\xi_r \in K(B(Z/p^r))$ be the image of $\xi \in K(CP^\infty)$; this element satisfies the relation

$$(\xi_r)^{p^r} = 1 .$$

LEMMA 6.5.1. $K\Lambda^0(B(Z/p^r))$ *has a* Λ*-base consisting of the elements* ξ_r^i, *where* i *runs over the residue classes* mod p^r; *and*

$$K\Lambda^1(B(Z/p^r)) = 0 .$$

I continue stating results before giving the proofs.

COROLLARY 6.5.2. *For each* r *and each element* $a \in A(\Lambda)$ *there is a finite* Λ*-linear combination*

$$b = \sum_k \lambda_k \Psi^k$$

of operations Ψ^k *which has the same effect as* a *in* $K(B(Z/p^r))$, *and indeed in* $K(B(Z/p^{r'}))$ *for all* $r' \leq r$.

LEMMA 6.5.3. *The maps*

$$\cdots \longrightarrow B(Z/_{p^r}) \longrightarrow B(Z/_{p^{r+1}}) \longrightarrow \cdots \longrightarrow BU(1) = CP^\infty$$

induce an isomorphism from $K\Lambda(CP^\infty)$ *to* $\varprojlim_r K\Lambda(B(Z/p^r))$.

COROLLARY 6.5.4. *The composite map*

$$A(\Lambda) \longrightarrow \varprojlim_r K\Lambda(B(Z/p^r)) .$$

is iso.

Proof of Lemma 6.5.1. We use the Atiyah-Hirzebruch spectral sequence. Let $R(Z/_{p^r})$ be the representation ring of $Z/_{p^r}$; it is Z-free, with the 1-dimensional representations of $Z/_{p^r}$ as free generators. We have an embedding of $R(Z/p^r)$ in $K^0(BZ/p^r))$, and this carries the 1-dimensional representations to the elements ξ_r^i (where i runs over the residue classes mod p^r). Here $K^0(B(Z/p^r))$ is filtered by the usual skeleton filtration; by restriction we get a filtration $R(Z/_{p^r})_*$ on $R(Z/_{p^r})$. The quotient

$$R(Z/_{p^r})_{2m} \Big/ R(Z/_{p^r})_{2m+2}$$

is Z for $m = 0$, $Z/_{p^r}$ for $m > 0$. We filter $\Lambda \otimes R(Z/_{p^r})$ by using the subgroups

$$\Lambda \otimes R(Z/_{p^r})_{2m} ;$$

we have then a filtration-preserving map

$$\Lambda \otimes R(Z/_{p^r}) \longrightarrow K\Lambda(B(Z/_{p^r})) .$$

Since Λ is torsion-free, $\Lambda \otimes$ preserves exactness; we have

$$\frac{\Lambda \otimes R(Z/_{p^r})_{2m}}{\Lambda \otimes R(Z/_{p^r})_{2m+2}} \cong \begin{cases} \Lambda & \text{for } m = 0 \\ \Lambda/p^r\Lambda & \text{for } m > 0 , \end{cases}$$

and this maps isomorphically to the E_2-term of the Atiyah-Hirzebruch spectral sequence. The Atiyah-Hirzebruch spectral sequence is trivial, and shows that $K\Lambda^1(B(Z/p^r))$ is zero, while $K\Lambda^0(B(Z/p^r))$ is the inverse limit of

$$\frac{\Lambda \otimes R(Z/p^r)}{\Lambda \otimes R(Z/p^r)_{2m}} ;$$

that is, we get the completion of $\Lambda \otimes R(Z/p^r)$ for the filtration topology. Now $R(Z/p^r)$ splits as the direct sum of Z (generated by 1) and the augmentation ideal $\tilde{R}(Z/p^r)$, which is free of rank p^r-1; similarly for $\Lambda \otimes R(Z/p^r)$. On the summand Λ of $\Lambda \otimes R(Z/p^r)$ the filtration is trivial, and on the summand $\Lambda \otimes \tilde{R}(Z/p^r)$ the filtration topology is the same as the p-adic topology. Since Λ is complete and Hausdorff for its p-adic topology, completing $\Lambda \otimes R(Z/p^r)$ with respect to the filtration topology gives $\Lambda \otimes R(Z/p^r)$ again. This proves Lemma 6.5.1.

Proof of Corollary 6.5.2. Suppose given $a \in A(\Lambda)$ and r. If

$$b = \sum_k \lambda_k \Psi^k ,$$

then

$$b(\xi_r) = \sum_k \lambda_k \xi_r^k ;$$

by Lemma 6.5.1 we can choose b so that

$$b(\xi_r) = a(\xi_r) .$$

There is a map

$$B(Z/_{p^r}) \longrightarrow B(Z/_{p^r})$$

which carries ξ_r to ξ_r^i, so by naturality we obtain

$$b(\xi_r^i) = a(\xi_r^i) .$$

Since both a and b are linear they have the same effect on any finite Z-linear combination

$$\sum_i \mu_i \xi_r^i .$$

Now such finite Z-linear combinations $\sum_i \mu_i \xi_r^i$ are dense in the filtration topology on $K(B(Z/_{p^r}))$, and both a and b are continuous with respect to this topology; so $ax = bx$ for every element $x \in K(B(Z/_{p^r}))$.

Finally, the assertion about $B(Z/_{p^{r'}})$ follows by naturality, since the map

$$K(B(Z/_{p^{r'}})) \longleftarrow K(B(Z/_{p^r}))$$

is epi. This proves Corollary 6.5.2.

Proof of Lemma 6.5.3. Let $Z/_{p^\infty}$ be the union of the subgroups $Z/_{p^r}$ of $U(1)$, and let us give $Z/_{p^\infty}$ the discrete topology, so that $B(Z/_{p^\infty})$ is an Eilenberg-MacLane space of type $(Z/_{p^\infty},1)$. Then $B(Z/_{p^\infty})$ may be constructed as the direct limit of spaces $B(Z/_{p^r})$. Since the inverse system

$$\cdots \longleftarrow K\Lambda^*(B(Z/_{p^r})) \longleftarrow K\Lambda^*(B(Z/_{p^{r+1}})) \longleftarrow \cdots$$

is a system of epimorphisms, the map

$$\varprojlim_r K\Lambda^*(B(Z/_{p^r})) \longleftarrow K\Lambda^*(B(Z/_{p^\infty}))$$

is iso. We consider now the map

$$B(Z/_{p^\infty}) \longrightarrow BU(1) .$$

The homology groups of this map are the relative homology groups when the map is converted into an inclusion, and the non-zero ones are isomorphic to $Z[^1/_p]$. Since Λ is Hausdorff for its p-adic topology we have

$$\mathrm{Hom}(Z[^1/_p], \Lambda) = 0 ;$$

and by calculation, using the fact that Λ is complete for its p-adic topology, we see that

$$\mathrm{Ext}(Z[^1/_p], \Lambda) = 0 .$$

So the ordinary universal coefficient theorem shows that the cohomology groups of the map are zero; that is,

$$H^*(B(Z/_{p^\infty}); \Lambda) \longleftarrow H^*(CP^\infty; \Lambda)$$

is iso. Now the Atiyah-Hirzebruch spectral sequence shows that

$$K\Lambda^*(B(Z/_{p^\infty})) \longleftarrow K\Lambda^*(CP^\infty)$$

is iso. This proves Lemma 6.5.3.

Corollary 6.5.4 follows from Lemmas 6.5.3 and 6.4.1.

For the real case, we can carry over Lemma 6.5.1 provided we replace the powers ξ_r^i by their underlying real 2-plane bundles $\eta_r^{(i)}$. We have to remember that

$$\eta_r^{(i)} = \eta_r^{(-i)}$$

as well as

$$\eta_r^{(i)} = \eta_r^{(j)} \text{ if } i \equiv j \bmod p^r ,$$

so it is necessary to run i over an appropriate set of representatives, say $0 \leq i \leq \frac{1}{2}p^r$. Since all these bundles are even-dimensional and orientable, we actually get a description of the subgroup

$$2\Lambda \oplus \widetilde{KSO}\Lambda(B(Z/_{p^r})) \subset KO\Lambda(B(Z/_{p^r})) .$$

Corollaries 6.5.2 and 6.5.4 carry over.

The remaining arguments of this section carry over to the real case without any really essential change, and I will mostly omit to mention this.

We can now use the isomorphism of Corollary 6.5.4 to define a direct-sum splitting of $A(\Lambda)$. With the notation of Lemma 6.5.1, let us define a direct-sum splitting of $K\Lambda(B(Z/_{p^r}))$ by

$$\sum_i \lambda_i \xi_r^i \mapsto \left(\sum_{i \not\equiv 0 \bmod p} \lambda_i \xi_r^i \right) + \left(\sum_{i \equiv 0 \bmod p} \lambda_i \xi_r^i \right) .$$

This decomposition clearly passes to the inverse limit, and defines a direct-sum splitting of $A(\Lambda)$, say

(6.5.5) $$a \mapsto a' + a'' .$$

Here a' is the part of a which can be defined using operations Ψ^k with k prime to p (and a well-defined limiting process), while a'' is the part of a which can be defined using operations Ψ^k with k divisible by p (and a well-defined limiting process). This inverse-sum splitting depends on the assumption that Λ is complete and Hausdorff for its p-adic topology; it would not work without some such assumption.

It is now clear that Proposition 6.3.5 will follow from the following two results.

LEMMA 6.5.6. *If* $a \in A(\Lambda)$ *and* $a\,\mathrm{Tr} = \mathrm{Tr}\,a$ *in all our coverings* $B(Z/_{p^r}) \to B(Z/_{p^{r+1}})$, *then* $a'' = 0$.

LEMMA 6.5.7. *If* $a \in A(\Lambda)$ *and* $a'' = 0$, *then* $a \in \mathrm{Im}\,\Omega^\infty$.

Lemma 6.5.7 will be proved in §6.6; it is the business of this section to prove Lemma 6.5.6. I proceed to describe the transfers involved.

LEMMA 6.5.8. *The transfer for the covering*

$$B(Z/_{p^r}) \longrightarrow B(Z/_{p^{r+1}})$$

carries ξ_r^i *to* $\sum_j \xi_{r+1}^j$, *where* j *runs over the* p *residue classes* mod p^{r+1} *which reduce to* i mod p^r.

Proof. We have the following commutative diagram of groups, in which the lower horizontal arrow is the obvious epimorphism.

$$\begin{array}{ccc} Z/_{p^r} & \longrightarrow & 1 \\ \big\downarrow & & \big\downarrow \\ Z/_{p^{r+1}} & \longrightarrow & Z/_p \end{array}$$

From this we obtain the following commutative diagram, in which the vertical arrows are coverings.

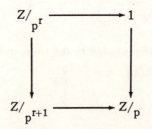

$$\begin{array}{ccc} B(Z/_{p^r}) & \longrightarrow & B(1) \\ \downarrow & & \downarrow \\ B(Z/_{p^{r+1}}) & \longrightarrow & B(Z/_p) \end{array}$$

For the right-hand covering, we use Atiyah's construction of the transfer in K-theory, and we see that

$$\mathrm{Tr}\, 1 = \rho ,$$

where ρ is the regular representation of $B(Z/_p)$. By naturality, we see that for the left-hand covering we have

$$\mathrm{Tr}\, 1 = 1 + \xi_{r+1}^{p^r} + \xi_{r+1}^{2p^r} + \cdots + \xi_{r+1}^{(p-1)p^r} .$$

The result for $\mathrm{Tr}(\xi_r^i)$ now follows using the multiplicative property of the transfer, because $\xi_r^i \in K(Z/_{p^r})$ is the image of $\xi_{r+1}^i \in K(Z/_{p^{r+1}})$.

Lemma 6.5.8 allows us to calculate how far our operations commute with transfer. Let the splitting

$$a \longmapsto a' + a''$$

be as in (6.5.5).

LEMMA 6.5.9. *The transfer for the covering*

$$B(Z/_{p^r}) \longrightarrow B(Z/_{p^{r+1}})$$

satisfies

$$(a\, \mathrm{Tr} - \mathrm{Tr}\, a)\, \xi_r^i = pa''\xi_{r+1}^i - \mathrm{Tr}\, a''\xi_r^i .$$

Proof. By naturality it is sufficient to prove the formula for $i = 1$. Then the formula is linear in a, and only depends on the value of a in

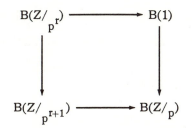

$B(Z/_{p^r})$ and $B(Z/_{p^{r+1}})$; so by Corollary 6.5.2 it is sufficient to check it for $a = \Psi^k$. First assume $k \equiv 0 \bmod p$, so that $a'' = \Psi^k = a$. Then

$$\begin{aligned}\Psi^k \operatorname{Tr} \xi_r &= \Psi^k\left(\sum_{\ell=0}^{p-1} \xi_{r+1}^{1+\ell p^r}\right) \\ &= \sum_{\ell=0}^{p-1} \xi_{r+1}^{k+k\ell p^r} \\ &= p\,\xi_{r+1}^{k} \quad (\text{since } kp^r \equiv 0 \bmod p^{r+1}) \\ &= p\,\Psi^k \xi_{r+1}\,.\end{aligned}$$

In the case $k \not\equiv 0 \bmod p$ we have $a'' = 0$, so that the right-hand side of the formula is zero; this case can be done the same way as the case $k \equiv 0 \bmod p$, but it is quicker to argue that $a = \Psi^k$ lies in $\operatorname{Im} \Omega^\infty$, so $a \operatorname{Tr} = \operatorname{Tr} a$. This proves the lemma.

By applying Lemma 6.5.9 to the two operations a' and a'' we can split it into two formulae

$$a' \operatorname{Tr} \xi_r^i = \operatorname{Tr} a' \xi_r^i$$

$$a'' \operatorname{Tr} \xi_r^i = pa'' \xi_{r+1}^i\,.$$

All these formulae remain true if ξ_{r+1}^i is replaced by any element $x \in K(B(Z/_{p^{r+1}}))$ and ξ_r^i is replaced by the image of x in $K(B/_{p^r}))$; we do not need to prove this.

Proof of Lemma 6.5.6. Suppose that $a \in A(\Lambda)$ commutes with transfer in all our coverings. As an initial argument, suppose that $a''(1) = \lambda 1$, where $\lambda \in \Lambda$. Using (6.5.9) for the covering

$$B(1) \longrightarrow B(Z/_p)\,,$$

and taking $i = 0$, we get

$$\begin{aligned}(a\,\mathrm{Tr} - \mathrm{Tr}\,a)1 &= pa''1 - \mathrm{Tr}\,a''1 \\ &= \lambda\left(p - \sum_{\ell=0}^{p-1} \xi_1^{\ell}\right),\end{aligned}$$

so we must have $\lambda = 0$. Suppose now as an inductive hypothesis that $a''\xi_r = 0$ in $B(Z/_{p^r})$; the induction starts with $r = 0$, by the argument above. We consider the covering

$$B(Z/_{p^r}) \longrightarrow B(Z/_{p^{r+1}});$$

using (6.5.9), we get

$$pa''\xi_{r+1} - \mathrm{Tr}\,a''\xi_r = 0 .$$

But here $\mathrm{Tr}\,a''\xi_r = 0$ by the inductive hypothesis, so we get $pa''\xi_{r+1} = 0$. Since $K\Lambda(B(Z/_{p^{r+1}}))$ is torsion-free by Lemma 6.5.1, we deduce that $a''\xi_{r+1} = 0$. This completes the induction, which shows that $a''\xi_r = 0$ for all r. Now Corollary 6.5.4 shows that $a'' = 0$, and this proves Lemma 6.5.6.

Proof of Proposition 6.3.7 from Proposition 6.4.5, assuming $\Lambda = Z_p^\wedge$. Again we have to make up for our neglect of SU.

Suppose given an additive cohomology operation

$$a\colon [W,BSU] \longrightarrow [W,BSU\Lambda]$$

where $\Lambda = Z_p^\wedge$. By [10], there is a unique additive cohomology operation b which makes the following diagram commutative.

$$\begin{array}{ccc} [W,BSU] & \xrightarrow{\ a\ } & [W,BSU\Lambda] \\ \downarrow & & \downarrow \\ [W,BU] & \xrightarrow{\ b\ } & [W,BU\Lambda] \end{array}$$

Now if a commutes with transfer in all our coverings, so does b; in fact, if $x \in [B(Z/p^r),BU]$, then $p^r x$ comes from some element $y \in [B(Z/p^r),BSU]$; so $a \operatorname{Tr} y = \operatorname{Tr} a y$ gives

$$b \operatorname{Tr} p^r x = \operatorname{Tr} b\, p^r x \ ,$$

and the result $b \operatorname{Tr} x = \operatorname{Tr} bx$ follows because $[B(Z/p^{r+1}),BU\Lambda]$ is torsion-free.

There are various additive operations c which make the following diagram commutative.

$$\begin{array}{ccc} [W,BU] & \xrightarrow{\ b\ } & [W,BU\Lambda] \\ \downarrow & & \downarrow \\ [W,Z\times BU] & \xrightarrow{\ c\ } & [W,\Lambda\times BU\Lambda] \end{array}$$

In fact, we have essentially only to pick the scalar λ in

$$c1 = \lambda 1 .$$

However, there is one and only one choice of c such that c commutes with transfer in all our coverings. More precisely, since c already commutes with transfer on elements of augmentation zero, it is necessary and sufficient to arrange that

$$c \operatorname{Tr} 1 = \operatorname{Tr} c1 ;$$

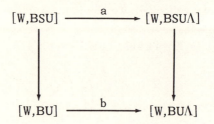

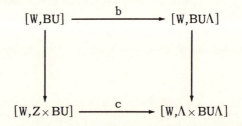

arguing as in the proof of (6.5.6), we see that for this it is necessary and sufficient to arrange that

$$c''1 = 0 .$$

Since the rather trivial operation Ψ^0 has $(\Psi^0)'' = \Psi^0$ there is one and only one choice of λ which secures $c''1 = 0$.

Assuming Proposition 6.3.5, there is a map

$$\mathbf{f}\colon \mathbf{K}(0,\cdots,\infty) \longrightarrow (\mathbf{K}(0,\cdots,\infty))\Lambda$$

such that $\Omega^\infty \mathbf{f} = c$. The map $\mathbf{f}$ determines a map

$$\mathbf{g}\colon \mathbf{K}(4,\cdots,\infty) \longrightarrow \mathbf{K}(4,\cdots,\infty))\Lambda \ .$$

In order to conclude that $\Omega^\infty \mathbf{g} = a$, it is sufficient to see that a is determined by the composite

$$\mathrm{BSU} \xrightarrow{a} \mathrm{BSU}\Lambda \longrightarrow \Lambda \times \mathrm{BU}\Lambda \ ;$$

and this is immediate, since $\mathrm{BSU}\Lambda$ is a retract of $\Lambda \times \mathrm{BU}\Lambda$. This completes the proof.

The argument which passes from b to c comes in handy in the real case also.

§6.6. *Slight convergence section*

The object of this section is to prove Lemma 6.5.7. I will give the argument for the complex case; the real case presents no essential difference.

First it is natural to discuss some topologies which can be put upon $[\mathbf{ku}, \mathbf{ku}\Lambda]$ and $A(\Lambda)$. The finest topology which can reasonably be put on $[\mathbf{ku}, \mathbf{ku}\Lambda]$ is the "filtration" topology, in which a typical neighborhood of zero consists of those maps $f\colon \mathbf{ku} \to \mathbf{ku}\Lambda$ which restrict to zero on the "skeleton" $\mathbf{ku}^n$; and indeed as a topologist, I probably have an *a priori* preference for this topology.

We also have to consider the composite map

$$[\mathbf{ku}, \mathbf{ku}\Lambda] \xrightarrow{\Omega^\infty} A(\Lambda) \longrightarrow K\Lambda(B(Z/_{p^r})) ,$$

and it will surely be desirable to have this continuous. For this purpose the obvious topology to put on $K\Lambda(B(Z/_{p^r}))$ is again the filtration topology, in which the neighborhood of zero are the kernels

$$\mathrm{Ker}\left[K\Lambda(B(Z/_{p^r})) \longrightarrow K\Lambda(B(Z/_{p^r})^n)\right] .$$

In fact, it would be unreasonable to hope that the composite map given above would be continuous for any finer topology on $K\Lambda(B(Z/_{p^r}))$. Now $K\Lambda(B(Z/_{p^r}))$ splits as the direct sum of a summand Λ (which presents no problem) and the subgroup of augmentation zero, $\widetilde{K\Lambda}(B(Z/_{p^r}))$; on the latter the filtration topology coincides with the p-adic topology, in which x is close to 0 if

$$x \equiv 0 \bmod p^n$$

for n sufficiently large. For algebraic purposes the p-adic topology is simple and convenient, so it becomes reasonable to consider the p-adic topology on the whole of $K\Lambda(B(Z/_{p^r}))$.

The obvious topology to put on

$$A(\Lambda) = \varprojlim_r K\Lambda(B(Z/_{p^r}))$$

is now the inverse-limit topology from the p-adic topologies on the terms $K\Lambda(B(Z/_{p^r}))$. I will call this the "limit" topology; this is the one most obviously related to the transfer calculations in §6.5. However, there are other definitions which appear equally reasonable, and which actually lead to the same topology.

First, recall from (6.4.1) the isomorphism

$$A(\Lambda) \xrightarrow{\cong} K\Lambda(CP^\infty) = \Lambda[[x]] .$$

We may topologize $\Lambda[[x]]$ by taking a typical neighborhood of zero to consist of those formal power-series $\sum_i \lambda_i x^i$ such that

$$\lambda_i \equiv 0 \bmod p^n \text{ for } 0 \le i \le m .$$

I will call this the "series" topology.

Secondly, we may topologize $A(\Lambda)$ by taking a typical neighborhood of zero to consist of those maps

$$a: Z \times BU \longrightarrow \Lambda \times BU\Lambda$$

such that

$$a_*: \pi_{2i}(Z \times BU) \longrightarrow \pi_{2i}(\Lambda \times BU\Lambda)$$

is congruent to zero mod p^n for $0 \le i \le m$. I will call this the "π_*"-topology; this is the one which is most clearly related to $[\mathbf{ku}, \mathbf{ku}\Lambda]$.

Of course, since we are dealing entirely with the K-theory of such simple spaces as $B(Z/_{p^r})$ and CP^∞, it must be elementary to relate these three topologies.

LEMMA 6.6.1. *These three topologies on* $A(\Lambda)$ *coincide.*

For our main purpose it is enough to know that the "limit" topology is finer than the "π_*"-topology; but it seems foolish not to know the full result.

I defer the proof of Lemma 6.6.1 until I have finished my explanations, and I turn to some topologies on $[\mathbf{ku}, \mathbf{ku}\Lambda]$. The first arises naturally from our discussion of $A(\Lambda)$; we take a typical neighborhood of zero to consist of those maps $\mathbf{f}: \mathbf{ku} \to \mathbf{ku}\Lambda$ such that

$$f_*: \pi_{2i}(ku) \longrightarrow \pi_{2i}(ku\Lambda)$$

is congruent to zero mod p^n for $0 \le i \le m$. In view of (6.3.4) and the discussion above, it amounts to topologizing $[ku, ku\Lambda]$ as a subgroup of $A(\Lambda)$, using the embedding Ω^∞.

The second topology is the p-adic topology which one would naturally put on the group

$$\mathrm{Hom}^*_{\pi_*(K)}(K_*(ku), \pi_*(K\Lambda))$$

in Lemma 6.4.7. To define a typical neighborhood of zero, we take a finite set of elements $x_1, x_2, \cdots, x_m$ in $K_*(ku)$, and define the typical neighborhood V of zero to be the set of those elements $f \in K\Lambda^*(ku)$ such that

$$\langle f, x_1 \rangle, \langle f, x_2 \rangle, \cdots, \langle f, x_m \rangle$$

are all congruent to zero mod p^n. I will call this the "K_*"-topology.

LEMMA 6.6.2. *The π_*-topology on $[ku, ku\Lambda]$ coincides with the K_*-topology.*

One must observe, of course, that both these topologies are strictly coarser than the "filtration topology" which I used to start this discussion.

I defer the proof of Lemma 6.6.2 also until I have finished my explanations. The justification of Lemmas 6.6.1 and 6.6.2 is that they allow one to use the following argument.

Proof of Lemma 6.5.7 from Lemmas 6.6.1, 6.6.2. Take an element a' in the subspace $a'' = 0$ of $A(\Lambda)$. We see from the definition of the splitting (6.5.5) that a' can be approximated by finite sums

$$a'_r = \sum_{k \not\equiv 0 \bmod p} \lambda_k \Psi^k ;$$

these sums a'_r lie in $\mathrm{Im}\,\Omega^\infty$, and tend to a' in the sense of the "limit" topology in $A(\Lambda)$. By Lemma 6.6.1 the "limit" topology is the same as the other topologies on $A(\Lambda)$. Let us identify $[\mathbf{ku}, \mathbf{ku}\Lambda]$ with $\mathrm{Im}\,\Omega^\infty$, using Ω^∞. It is clear from Lemma 6.4.7 that $[\mathbf{ku}, \mathbf{ku}\Lambda]$ is complete for its K_*-topology, and by Lemma 6.6.2 the K_*-topology is the same as the topology which $\mathrm{Im}\,\Omega^\infty$ has as a subspace of $A(\Lambda)$. We conclude that the sequence a'_r tends to a limit in $\mathrm{Im}\,\Omega^\infty$, that is, $a' \epsilon \mathrm{Im}\,\Omega^\infty$. This proves Lemma 6.5.7.

Proof of Lemma 6.6.2. First take V to be a neighborhood in the π_*-topology, consisting of those $\mathbf{f}$ such that

$$\mathbf{f}_*\colon \pi_{2i}(\mathbf{ku}) \longrightarrow \pi_{2i}(\mathbf{ku}\Lambda)$$

is congruent to $0 \bmod p^n$ for $0 \le i \le m$. It is equivalent to ask that

$$<\mathbf{f}, v^i> \equiv 0 \bmod p^n \text{ for } 0 \le i \le m ,$$

so that V is also a neighborhood in the K_*-topology.

Conversely, take V to be a neighborhood in the K_*-topology; it is sufficient to consider the case of a neighborhood defined by a single element $x \epsilon K_*(\mathbf{ku})$, so that it consists of those $\mathbf{f} \epsilon K\Lambda^*(\mathbf{ku})$ such that $<\mathbf{f}, x> \equiv 0 \bmod p^n$. The element x can be written as a finite linear combination

$$\sum_{r,s} \mu_{r,s}\, u^r v^s$$

with $\mu_{r,s} \epsilon Q$ and $s \ge 0$; we may suppose that it only contains terms with $0 \le s \le m$, and that by using a common denominator we write

$$p^t x = \sum_{r,s} \nu_{r,s}\, u^r v^s$$

with $\nu_{r,s} \epsilon Z_{(p)}$. We can specify a neighborhood V' in the π^*-topology by the congruences

$$\langle f, v^s \rangle \equiv 0 \bmod p^{n+t} \quad \text{for } 0 \leq s \leq m ;$$

then $f \in V'$ implies $f \in V$, i.e. $V' \subset V$. This proves Lemma 6.6.2.

Proof of Lemma 6.6.1. First I show that the "series" topology coincides with the "π_*"-topology.

Recall from the proof of Corollary 6.4.2 that the coefficients in the series

$$a(\xi) = \sum_i \lambda_i x^i$$

are related to the induced maps of homotopy as follows:

$$a_*: \pi_{2j}(Z \times BU) \longrightarrow \pi_{2j}(\Lambda \times BU\Lambda)$$

is multiplication by $\mu_j = \sum_{i=0} \lambda_i \beta_{ij}$, where $\beta_{ij} \in Z$. So if we subject the λ's to the congruences

$$\lambda_i \equiv 0 \bmod p^n \quad \text{for } 0 \leq i \leq m ,$$

that ensures

$$\mu_j \equiv 0 \bmod p^n \quad \text{for } 0 \leq i \leq m .$$

Conversely, let p^t be the p-primary factor of

$$\det_{\substack{0 \leq i \leq m \\ 0 \leq j \leq m}} (\beta_{ij}) = 1!\, 2!\, 3! \cdots m! ;$$

if we subject the μ's to the congruences

$$\mu_j \equiv 0 \bmod p^{n+t} \quad \text{for } 0 \leq j \leq m ,$$

that ensures

$$\lambda_j \equiv 0 \bmod p^n \quad \text{for} \quad 0 \leq i \leq m .$$

This shows that the "series" topology coincides with the "π_*"-topology.

Secondly I show that the "series" topology is finer than the "limit" topology. Suppose given a neighborhood V of zero in the limit topology, consisting say of elements $a \in A(\Lambda)$ such that

$$a(\xi_r) \equiv 0 \bmod p^n \quad \text{in} \quad K\Lambda(B(Z/_{p^r})) .$$

In $K(B(Z/_{p^r}))$ we have an equation

$$(\xi_r)^{p^r} = 1 ,$$

from which we deduce

$$(x_r)^{p^r} = (\xi_r - 1)^{p^r} = py$$

for some suitable y. Taking powers, we get

$$(x_r)^{np^r} = p^n y^n .$$

So if

$$b(\xi) = \sum_{i=m}^{\infty} \lambda_i x^i$$

where $m = np^r$, this will ensure

$$b(\xi_r) = p^n z$$

for some suitable z. Let us write

$$a(\xi) = \sum_{i=0}^{\infty} \lambda_i x^i ;$$

then the congruences

$$\lambda_i \equiv 0 \bmod p^n \quad \text{for} \quad 0 \leq i < m = np^r$$

define a neighborhood of zero in the "series" topology, and ensure that a lies in V.

Thirdly I show that the "limit" topology is finer than the "series" topology. Suppose given a neighborhood V of zero in the series topology, say consisting of those elements a such that

$$a(\xi) = \sum_i \lambda_i x^i$$

with

$$\lambda_i \equiv 0 \bmod p^n \quad \text{for} \quad i \leq m .$$

We will construct by downwards induction over k a number $r = r(k)$ with the following property: if

$$b(\xi) = \sum_{i \geq k} \mu_i x^i$$

and $b(\xi_r)$ maps to zero in $K\Lambda(B(Z/_{p^r})^{2m})$, then $b \in V$. The induction starts trivially with $k = m+1$, or with $k = m$ by taking $r = n$. Suppose then that $r = r(k+1)$ is constructed, where $1 \leq k \leq m$. Since $\widetilde{K}(B(Z/_{p^r})^{2m})$ is annihilated by p^{rm}, we can find a power p^s of p such that the element $(\xi_r - 1)^k = (x_r)^k$ is annihilated by p^s in $\widetilde{K}(B(Z/_{p^r})^{2m})$. Take now

$$r' = r(k) = \operatorname{Max}(n, r, s) .$$

If

$$b(\xi) = \sum_{i \geq k} \mu_i x^i$$

and $b(\xi_{r'})$ maps to zero in $K\Lambda(B(Z/_{p^{r'}})^{2m})$, then by considering the term of lowest filtration we see that

$$\mu_k \equiv 0 \bmod p^{r'} .$$

Now we can consider the operation c with

$$c(\xi) = \sum_{i \geq k+1} \mu_i x^i ;$$

it has the property that $c(\xi_r)$ maps to zero in $K\Lambda(B(Z/_{p^r})^{2m})$; so c lies in V, by the inductive hypothesis. It follows that b lies in V. This completes the induction.

The induction finishes with $k = 1$, and constructs a number r such that if

$$b(\xi) = \sum_{i \geq 1} \mu_i x^i$$

and $b(\xi_r)$ maps to zero in $K\Lambda(B(Z/_{p^r})^{2m})$, then $b \in V$.

As a final step, choose s so that p^s annihilates $\widetilde{K\Lambda}(B(Z/_{p^r})^{2m})$; for example, $s = rm$ will do. Let $N = \max(s,n)$, and let V' be the set of $a \in A(\Lambda)$ such that $a(\xi_r)$ is divisible by p^N in $K\Lambda(B(Z/_{p^r}))$. Then V' is a neighborhood of zero in the limit topology, and $V' \subset V$.

This completes the proof of Lemma 6.6.1, which completes the proof of (6.5.7) and (6.3.5).

The reader who feels that the third part of the proof of (6.6.1) is unnecessarily wasteful is urged to study the non-trivial extensions in $\widetilde{K}(B(Z/_{p^r})^{2m})$ for particular values of $r > 1$ and $m > 1$.

CHAPTER 7

THE STATE OF THE ART

§7.1. *Survey*

In this chapter I will make some brief and probably wholly inadequate attempt to give an impression of the current state of this subject.

It seems best to begin by continuing the exposition which I began in §1.8. I have said that from the point of view of the topology of manifolds, one of the things that topologists are required to do is to investigate the following sequence of groups and homomorphisms.

$$K_O(X) \to K_{PL}(X) \to K_{Top}(X) \to K_F(X) .$$

Alternatively, we may say that we wish to take the representing spaces

$$BO, \quad BPL, \quad BTop, \quad BF$$

together with the corresponding coset spaces, and analyze or decompose them as infinite loop spaces in terms of more elementary parts. Of course we would like not only an analysis of objects, but also an analysis of maps; ideally one would like a list of maps in which each geometrically important map between these objects occurs once and only once, together with an index annotating all the interesting maps constructed by earlier authors and showing where each one occurs in the master list. When I originally gave these lectures, I felt that such a degree of systematization of the subject was far from achieved. Today (9/9/77) it is well under way; and the reader may consult [99] for evidence. For example, it has been found possible to prove that a map constructed by "discrete models" agrees with a map constructed by other methods.

Let me return to the project of an "analysis of objects." We have said in Chapter 5 that the homotopy groups of BF are the stable homotopy groups of spheres, and therefore they are complicated. However, if we localize at an odd prime, the stable homotopy groups of spheres split as a direct sum, in the form

$$\pi_n^S(S^0) \cong (\mathrm{Im}\ J)_n \oplus (\mathrm{Coker}\ J)_n .$$

This may be shown starting from the results surveyed in Chapter 5. Therefore, we reconcile ourselves with regret to the prospect of admitting one space or spectrum whose homotopy groups are Coker J; and to this space or spectrum we consign all the unsolved problems of homotopy theory. Any space or spectrum so conceived and dedicated may be called Coker J. Of course, subscripts, superscripts and other decoration may be affixed by those authors who in an excess of zeal conceive and dedicate more than one such object, or view the same object in more than one way.

A slight anomaly occurs if we localize at the prime 2; it remains true that $\pi_n^S(S^0)$ splits as the sum of a "known part" related to K-theory and an "unknown part," but the description of the "known part" has to be slightly different; besides the image of the classical J-homomorphism, it should contain certain $Z/2$ subgroups generated by the elements μ_r for $r \equiv 1,2 \bmod 8$ [5]; these are things which homotopy-theorists know and love, but which need not concern anyone else.

Returning again to the project of an "analysis of objects," we admit the deplorable necessity of one rogue object Coker J. All the remaining parts of our decomposition we would like to keep firmly under control; and in practice so far this means that they must be closely related to the known spectra which represent versions of connective K-theory. (When I say "closely related," one would admit constructions such as cofibrations starting from a known map between known K-theory spectra.) It follows that the sort of result set out in Chapter 6 offers good hope that the maps between such known spectra can be kept under control. On this point, see

[69] Theorem 6.1 (though I distrust some of the p-adic results in this manuscript), or [70] Theorem 9.3 p. 232, Theorem 9.9 p. 236.

One can get a good idea of the progress in this field by comparing the problems in [95] with the results in [99]. I may divide the concrete problems of the sort I am now considering into problems at three levels.

At the first or easiest level, May [95] states as Problem 6, "When is an H-map between two infinite loop spaces, both equivalent to BSO localized or completed at p, an infinite loop map?"

As we have seen, this is exactly the problem which Madsen, Snaith and Tornehave solved; and as I have suggested, we are now not too much afraid of other problems at this level.

I pass on to state two problems at the second or intermediate level.

QUESTION 7.1.1. Atiyah, Bott and Shapiro [19] provide a KO-orientation for Spin-bundles, that is, a natural transformation

$$K_{\mathrm{Spin}}(X) \xrightarrow{\sigma} K_{\mathrm{Spin;KO}}(X)$$

such that $\pi\sigma = 1$, where $K_{\mathrm{Spin;KO}}(X)$ is a Grothendieck group defined in terms of Spin-bundles with given KO-orientations, and

$$\pi\colon K_{\mathrm{Spin;KO}}(X) \longrightarrow K_{\mathrm{Spin}}(X)$$

is defined by forgetting the orientation. Does σ extend to a natural transformation $\boldsymbol{\sigma}$ of cohomology theories such that $\boldsymbol{\pi\sigma} = \mathbf{1}$?

QUESTION 7.1.2. Similar to (7.1.1), but with the orientation due to Atiyah, Bott and Shapiro replaced by Sullivan's orientation of STop-bundles over KOZ[½] (see [149] §6).

These two questions are Conjectures 2 and 3 of [95]. Affirmative answers have been obtained modulo a slight modification of the questions; the modified questions are sufficient for the purposes envisaged when the original questions were asked. The modification is as follows: the equation analogous to $\pi\sigma = 1$ in (7.1.1) takes place not in $K_{Spin;KO}(X)$, but in the group $K_{SF}(X)$ defined in terms of spherical fibrations rather than Spin-bundles (see §1.8). For these results, see [99] Theorem 7.11 p. 135, Theorem 7.16 p. 137.

I pass on to state the only problem at the third or deepest level, and this concerns "infinite-loop versions" of the Adams conjecture discussed in Chapter 5.

First I state a complex version of the conjecture. Let us accept that the map

$$\widetilde{K}(X) \longrightarrow \widetilde{K}_{SF}(X)$$

extends to a natural transformation of cohomology theories, corresponding to a map of spectra

$$\mathbf{bu} \longrightarrow \mathbf{BSF}\ .$$

QUESTION 7.1.3. Is the composite

$$\mathbf{bu} \xrightarrow{\psi^k - 1} \mathbf{bu}Z[1/k] \longrightarrow \mathbf{BSF}Z[1/k]$$

nullhomotopic as a map of spectra?

This is conjecture 1 of [95]. It has recently been answered in the affirmative by two independent pieces of work, due to E. M. Friedlander [61] and R. F. Seymour [132]. Friedlander's methods are based on Sullivan's method of étale homotopy in algebraic geometry; they involve adjusting that method and the machinery of Chapter 2 till they fit. Seymour's methods involve the geometric construction of the cohomology theory represented by the fibre of $(\Psi^k - 1)$, as in [131].

QUESTION 7.1.4. What should be the "real" analogue of Question 7.1.3 at the prime $p = 2$?

At an odd prime p there is no trouble; the obvious real analogue of (7.1.3) is answered by the answer to (7.1.3). The prime $p = 2$ is unavoidably different. It is easy to formulate a statement so weak as to be unhelpful (for example, empty) or so strong as to be untrue (for example, the obvious real analogue of (7.1.3)). The question asks for the "right" statement. It appears that a suitable candidate has been found, but I have not yet heard that there is good reason to hope for a proof of it.

I note also another recent advance; it has been proved that the different machines in Chapter 2 are essentially equivalent. More precisely, a compulsively reasonable set of axioms is given on a "machine" which converts space-level input into spectrum-level output; it is shown that a suitable version of May's machine satisfies the axioms; and it is shown that any machine satisfying the axioms is equivalent to Segal's machine, which is found convenient for comparison. As happened with the Eilenberg-Steenrod axioms, the onus is probably now on any builder of a new machine to check that his machine satisfies the axioms. This advance is due to May and Thomason [100]. The basic idea comes from work of Fiedorowicz which proves the uniqueness of the spectra of algebraic K-theory [58].

A supplementary result shows the uniqueness of machines which pass from "permutative category" input to spectrum-level output.

It seems then that the theory has reached a reasonably satisfactory state, and is close to completing its main aims. On this cheerful note I end my survey.

REFERENCES

[1] J. F. Adams, Vector fields on spheres, Ann. of Math.(2) 75 (1962), 603-632.
MR 25 (1963) 2614.

[2] __________, On the groups J(X). I. Topology 2 (1963), 181-195.
MR 28 (1964) 2553.

[3] __________, On the groups J(X). II. Topology 3 (1965), 131-171.
MR 33 (1967) 6626.

[4] __________, On the groups J(X). III. Topology 3 (1965), 193-222.
MR 33 (1967) 6627.

[5] __________, On the groups J(X). IV. Topology 5 (1966), 21-71.
MR 33 (1967) 6628.
With a correction, Topology 7 (1968), 331.
MR 37 (1969) 5874.

[6] __________, Lectures on generalised cohomology, in Lecture Notes in Mathematics vol. 99, Springer 1969, 1-138.
MR 40 (1970) 4943.

[7] __________, Algebraic topology in the last decade, Proceedings of Symposia in Pure Mathematics 22, Amer. Math. Soc. 1971, 1-22.
MR 47 (1974) 5858.

[8] __________, The Kahn-Priddy theorem, Proc. Cambridge Philos. Soc. 73 (1973) 45-55.
MR 46 (1973) 9976.

[9] __________, Stable homotopy and generalised cohomology, University of Chicago Press 1974.
MR 53 (1977) 6534.

[10] __________, Primitive elements in the K-theory of BSU, Quarterly Jour. of Math. 27 (1976) 253-262.

[11] J. F. Adams and F. W. Clarke, Stable operations on complex K-theory, to appear in the Illinois Journal of Mathematics.

[12] J. F. Adams, A. S. Harris and R. M. Switzer, Hopf algebras of cooperations for real and complex K-theory, Proc. London Math. Soc. (3) 23 (1971) 385-408.
MR 45 (1973) 2694.

[13] J. F. Adams and P. Hoffman, Operations on K-theory of torsion-free spaces, Math. Proc. Cambridge Philos. Soc. 79 (1976) 483-491.
MR 53 (1977) 4063.

[14] J. F. Adams and S. B. Priddy, Uniqueness of BSO, Math. Proc. Cambridge Philos. Soc. 80 (1976) 475-509.

[15] J. F. Adams and G. Walker, On complex Stiefel manifolds, Proc. Cambridge Philos. Soc. 61 (1965) 81-103.
MR 30 (1965) 1516.

[16] D. W. Anderson, Spectra and Γ-sets, in Proceedings of Symposia in Pure Mathematics 22, Amer. Math. Soc. 1971, 23-30.
MR 51 (1976) 4232.

[17] M. F. Atiyah, Characters and cohomology of finite groups, Publ. Math. of the I.H.E.S. no.9 (1961) 23-64.
MR 26 (1963) 6228.

[18] M. F. Atiyah and R. Bott, On the periodicity theorem for complex vector bundles, Acta Math. 112 (1964) 229-247.
MR 31 (1966) 2727.

[19] M. F. Atiyah, R. Bott and A. Shapiro, Clifford modules, Topology 3 Suppl. 1 (1964) 3-38.
MR 29 (1965) 5250.

[20] M. F. Atiyah and G. Segal, Equivariant K-theory and completion, Jour. Differential Geometry 3 (1969) 1-18.
MR 41 (1971) 4575.

[21] M. G. Barratt, A note on the cohomology of semigroups, Jour. London Math. Soc. 36 (1961) 496-498.
MR 25 (1963) 3078.

[22] ――――――, A free group functor for stable homotopy, in Proceedings of Symposia in Pure Mathematics 22, Amer. Math. Soc. 1971, 31-35.
MR 48 (1974) 3043.

[23] M. G. Barratt, On Γ-structures, Bull. Amer. Math. Soc. 77 (1971) 1099. MR 46 (1973) 6352.

[24] M. G. Barratt and P. J. Eccles, Γ^+-structures. I. A free group functor for stable homotopy theory. Topology 13 (1974) 23-45. MR 50 (1975) 1234a.

[25] ________, Γ^+-structures. II. A recognition principle for infinite loop spaces. Topology 13 (1974) 113-126. MR 50 (1975) 1234b.

[26] ________, Γ^+-structures. III. The stable structure of $\Omega^\infty\Sigma^\infty A$. Topology 13 (1974) 199-207. MR 50 (1975) 1234c.

[27] M. G. Barratt and S. B. Priddy, On the homology of nonconnected monoids and their associated groups, Comment. Math. Helv. 47 (1972) 1-14. MR 47 (1974) 3489.

[28] J. Beck, On H-spaces and infinite loop spaces, in Lecture Notes in Mathematics no.99, Springer 1969, 139-153. MR 40 (1970) 2079.

[29] ________, Classifying spaces for homotopy-everything H-spaces, in Lecture Notes in Mathematics no.196, Springer 1971, 54-62. MR 45 (1973) 1160.

[30] J. C. Becker, Characteristic Classes and K-theory, in Lecture Notes in Mathematics no.428, Springer 1974, 132-143. MR 51 (1976) 14046.

[31] J. C. Becker, A. Casson and D. H. Gottlieb, The Lefschetz number and fibre preserving maps, Bull. Amer. Math. Soc. 81 (1975) 425-427. MR 53 (1977) 14474.

[32] J. C. Becker and D. H. Gottlieb, Applications of the evaluation map and transfer map theorems, Math. Ann. 211 (1974) 277-288. MR 50 (1975) 11231.

[33] ________, The transfer map and fiber bundles, Topology 14 (1975) 1-12. MR 51 (1976) 14042.

[34] J. M. Boardman, Thesis, Cambridge 1964.

[35] J. M. Boardman, Stable homotopy theory, mimeographed notes, University of Warwick, 1966.

[36] ―――――, Stable homotopy theory, mimeographed notes, Johns Hopkins, 1969/70.

[37] ―――――, Monoids, H-spaces and tree surgery, mimeographed notes, Haverford College, 1969.

[38] ―――――, Homotopy structures and the language of trees*, Proceedings of Symposia in Pure Mathematics 22, Amer. Math. Soc. 1971, 37-58.
MR 50 (1975) 3215.

[39] J. M. Boardman and R. M. Vogt, Homotopy-everything H-spaces, Bull. Amer. Math. Soc. 74 (1968) 1117-1122.
MR 38 (1969) 5215.

[40] ―――――, Homotopy invariant algebraic structures on topological spaces, Lecture Notes in Mathematics no.347, Springer 1973.
MR 54 (1977) 8623a.

[41] A. Borel and F. Hirzebruch, Characteristic classes and homogeneous spaces. I. Amer. Jour. Math. 80 (1958) 458-538.
MR 21 (1960) 1586.

[42] R. Bott, The stable homotopy of the classical groups, Proc. Nat. Acad. Sci. USA 43 (1957) 933-935.
MR 21 (1960) 1588.

[43] ―――――, The stable homotopy of the classical groups, Annals of Math.(2) 70 (1959) 313-317.
MR 22 (1961) A987.

[44] R. Bott and H. Samelson, On the Pontryagin product in spaces of paths, Comment. Math. Helv. 27 (1953) 320-337.
MR 15 (1954) p. 643.

[45] A. K. Bousfield and D. M. Kan, Localisation and completion in homotopy theory, Bull. Amer. Math. Soc. 77 (1971) 1006-1010.
MR 45 (1973) 5994.

[46] ―――――, Homotopy limits, completions and localisations, Lecture Notes in Mathematics no.304, Springer 1972.
MR 51 (1976) 1825.

*They whisper in the wind.

[47] W. Browder, Homology operations and loop spaces, Illinois Jour. Math. 4 (1960) 347-357.
MR 22 (1961) 11395.

[48] E. H. Brown, Cohomology theories, Annals of Math.(2) 75 (1962), 467-484.
MR 25 (1963) 1551.
With a correction, Annals of Math.(2) 78 (1963) 201.
MR 27 (1964) 749.

[49] ———, Abstract Homotopy Theory, Trans. Amer. Math. Soc. 119 (1965) 79-85.
MR 32 (1966) 452.

[50] H. Cartan, Séminaire H. Cartan 12, 1959/60. Periodicité des groupes d'homotopie stables des groupes classiques, d'apres Bott. Sécretariat mathématique, 11 rue Pierre Curie, Paris 5e. 1961.
MR 28 (1964) 1092.

[51] H. Cartan and S. Eilenberg, Homological Algebra, Princeton Univ. Press 1956.
MR 17 (1956) p. 1040.

[52] P. V. Z. Cobb, P_n-spaces and n-fold loop spaces, Bull. Amer. Math. Soc. 80 (1974) 910-914.
MR 50 (1975) 14737.

[53] F. R. Cohen, T. J. Lada and J. P. May, The homology of iterated loop spaces, Lecture Notes in Mathematics no. 533, Springer 1976.

[54] A. Dold and R. K. Lashof, Principal quasifibrations and fibre homotopy equivalence of bundles. Illinois Jour. Math. 3 (1959) 285-305.
MR 21 (1960) 331.

[55] E. Dyer and R. K. Lashof, Homology of iterated loop spaces, Amer. Jour. Math. 84 (1962) pp. 35-38.
MR 25 (1963) 4523.

[56] B. Eckmann, On complexes with operators, Proc. Nat. Acad. Sci. USA 39 (1953) 35-42.
MR 15 (1954) p. 459.

[57] S. Eilenberg and S. MacLane, On the groups $H(\pi,n)$. I. Annals of Math.(2) 58 (1953) 55-106.
MR 15 (1954) p. 54.

[58] Z. Fiedorowicz, A note on the spectra of algebraic K-theory, to appear in Topology.

[59] H. Freudenthal, Uber die Klassen von Spharenabbildungen I, Composito Math. 5 (1937) pp. 299-314.

[60] E. M. Friedlander, Fibrations in etale homotopy theory, Publ. Math. of the I.H.E.S. 42 (1973) 5-46.
MR 51 (1976) 3175.

[61] ———, Stable Adams' conjecture, submitted to the Math. Proc. Camb. Philos. Soc.

[62] S. M. Gersten, On the spectrum of algebraic K-theory, Bull. Amer. Math. Soc. 78 (1972) 216-219.
MR 45 (1973) 8705.

[63] P. J. Hilton, On the homotopy groups of the union of spheres, J. London Math. Soc. 30 (1955) 154-172.
MR 16 (1955) p. 847.

[64] P. J. Hilton, G. Mislin and J. Roitberg, Topological localisation and nilpotent groups, Bull. Amer. Math. Soc. 78 (1972) 1060-1063.
MR 47 (1974) 1066.

[65] ———, Homotopical localisation, Proc. London Math. Soc.(3) 26 (1973) 693-706.
MR 48 (1974) 5063.

[66] ———, Localisation of nilpotent groups and spaces, North-Holland 1975.

[67] P. J. Hilton and J. Roitberg, Note on principal S^3-bundles. Bull. Amer. Math. Soc. 74 (1968) 957-959.
MR 37 (1969) 5889.

[68] ———, On principal S^3-bundles over spheres, Annals of Math. (2) 90 (1969) 91-107.
MR 39 (1970) 7624.

[69] L. Hodgkin and V. P. Snaith, The K-theory of some more well-known spaces, preprint.

[70] ———, Topics in K-theory, Lecture Notes in Mathematics no 496. Springer 1975.

[71] D. Husemoller, Fibre Bundles, McGraw-Hill 1966.
MR 37 (1969) 4821.

[72] J. R. Isbell, On coherent algebras and strict algebras, J. Algebra 13 (1969) 299-307.
MR 40 (1970) 2729.

[73] I. M. James, Reduced product spaces, Annals of Math.(2) 62 (1955) 170-197.
MR 17 (1956) p. 396.

[74] D. S. Kahn and S. B. Priddy, Applications of the transfer to stable homotopy theory, Bull. Amer. Math. Soc. 78 (1972) 981-987.
MR 46 (1973) 8220.

[75] D. M. Kan, Adjoint functors, Trans. Amer. Math. Soc. 87 (1958) 294-329.
MR 24 (1962) A1301.

[76] D. M. Kan and W. P. Thurston, Every connected space has the homology of a $K(\pi,1)$, Topology 15 (1976) 253-258.
MR 54 (1977) 1210.

[77] T. Kudo and S. Araki, On $H^*(\Omega^N(S^n);Z_2)$. Proc. Japan Acad. 32 (1956) 333-335.
MR 18 (1957) p. 143.

[78] ————, Topology of H_n-spaces and H-squaring operations, Mem. Fac. Sci. Kyusyu Univ. Ser. A 10 (1956) 85-120.
MR 19 (1958) p. 442.

[79] J. M. Lemaire, Le transfert dans les espaces fibrés (d'apres J. Becker et D. Gottlieb). Séminaire N. Bourbaki no.472, 1975.

[80] H. J. Ligaard, Infinite loop maps from SF to $BO_\otimes$ at the prime 2, to appear in the Illinois Journal of Math.

[81] E. L. Lima, The Spanier-Whitehead duality in new homotopy categories, Summa Brasiliensis Math. 4 (1959) 91-148.
MR 22 (1961) 7121.

[82] ————, Stable Postnikov invariants and their duals, Summa Brasiliensis Math. 4 (1960) 193-251.
MR 26 (1963) 772.

[83] S. MacLane, Categorical Algebra, Bull. Amer. Math. Soc. 71 (1965) 40-106.
MR 30 (1965) 2053.

[84] ———————, Categories for the working mathematician, Graduate texts in mathematics no.5, Springer 1971.
MR 50 (1975) 7275.

[85] J. MacNab, Categories for the idle mathematician; all you need to know, Proceedings of the Philharmonic Society of Zanzibar 17 (1976) 10-9.

[86] I. Madsen, V. P. Snaith and J. Tornehave, Homomorphisms of spectra and bundle theories, preprint, Aarhus 1974/75.

[87] ———————, Infinite loop maps in geometric topology, Math. Proc. Camb. Phil. Soc. 81 (1977) 399-430.

[88] J. P. May, Simplicial objects in algebraic topology, Van Nostrand Mathematical Studies no.11, Van Nostrand 1967.
MR 36 (1968) 5942.

[89] ———————, Categories of spectra and infinite loop spaces, in Lecture Notes in Mathematics no.99, Springer 1969, 448-479.
MR 40 (1970) 2073.

[90] ———————, A general algebraic approach to Steenrod operations, in Lecture Notes in Mathematics no.168, Springer 1970, 153-231.
MR 43 (1972) 6915.

[91] ———————, Homology operations on infinite loop spaces, in Proceedings of Symposia in Pure Mathematics 22, Amer. Math. Soc. 1971, 171-185.
MR 47 (1974) 7740.

[92] ———————, The geometry of iterated loop spaces, Lecture Notes in Mathematics no.271, Springer 1972.

[93] ———————, E_∞ spaces, group completions, and permutative categories, in "New developments in topology", London Math. Soc. lecture notes no.11, Cambridge Univ. Press 1974, 61-93.
MR 49 (1975) 3915.

[94] ———————, Classifying spaces and fibrations. Mem. Amer. Math. Soc. 155 (1975).
MR 51 (1976) 6806.

[95] J. P. May, Problems in infinite loop space theory, in Notas de matematica y symposia Vol. 1, Sociedad Matematica Mexicana 1975, 106-125.

[96] ———, Infinite loop space theory. Bull. Amer. Math. Soc. 83 (1977) 456-494.

[97] ———, The homotopical foundations of algebraic topology. To appear as a Monograph of the London Mathematical Society.

[98] ———, H_∞ ring spectra, in the proceedings of the topology conference in Stanford, 1976, to appear in the series Proceedings of Symposia in Pure Math., Amer. Math. Soc.

[99] J. P. May, F. Quinn, N. Ray and J. Tornehave, E_∞ ring spaces and E_∞ ring spectra, Lecture Notes in Math. no.577, Springer 1977.

[100] J. P. May and R. Thomason, The uniqueness of infinite loop space machines, to appear.

[101] D. McDuff and G. Segal, Homology fibrations and the "group completion" theorem, Invent. Math. 31 (1976) 279-284.
MR 53 (1977) 6547.

[102] R. J. Milgram, Iterated loop spaces, Annals of Math. 84 (1966) 386-403.
MR 34 (1967) 6767.

[103] ———, The bar construction and abelian H-spaces, Illinois Jour. Math. 11 (1967) 242-250.
MR 34 (1967) 8404.

[104] J. W. Milnor, On the cobordism ring Ω^* and a complex analogue. I. Amer. Jour. Math. 82 (1960) 505-521.
MR 22 (1961) 9975.

[105] ———, Microbundles and differentiable structures, mimeographed notes, Princeton 1961.

[106] ———, On axiomatic homology theory, Pacific Jour. Math. 12 (1962) 337-341.
MR 28 (1964) 2544.

[107] ———, Topological manifolds and smooth manifolds, in Proc. Intern. Congress Math. 1962, Institut Mittag-Leffler 1963, 132-138.
MR 28 (1964) 4553a.

[108] J. W. Milnor, Microbundles, Topology 3 Suppl. 1 (1964) 53-80. MR 28 (1964) 4553b.

[109] M. Mimura, G. Nishida and H. Toda, Localisation of CW-complexes and its applications, J. Math. Soc. Japan. 23 (1971) 593-624. MR 45 (1973) 4413.

[110] C. Morlet, Microfibrées et structures différentiables, Séminaire Bourbaki 1963/64, Fasc. 1, exposé 263. MR 33 (1967) 729.

[111] M. Morse, The calculus of variations in the large, Amer. Math. Soc. Colloquium Publications vol. 18, Amer. Math. Soc. 1934.

[112] G. Nishida, Cohomology operations in iterated loop spaces, Proc. Japan Acad. 44 (1968) 104-109. MR 39 (1970) 2156.

[113] ———, The nilpotency of elements of the stable homotopy groups of spheres, J. Math. Soc. Japan 25 (1973) 707-732. MR 49 (1975) 6236.

[114] S. B. Priddy, On $\Omega^\infty S^\infty$ and the infinite symmetric group, in Proceedings of Symposia in Pure Math. 22, Amer. Math. Soc. 1971, 217-220. MR 50 (1975) 11226.

[115] ———, Transfer, symmetric groups, and stable homotopy theory, in Lecture Notes in Mathematics no.341, Springer 1973, 244-255. MR 50 (1975) 3219.

[116] D. G. Quillen, Some remarks on étale homotopy theory and a conjecture of Adams, Topology 7 (1968) 111-116. MR 37 (1969) 3572.

[117] ———, Cohomology of Groups, in Proceedings of the International Congress of Mathematicians 1970, Gauthier-Villars 1971, vol. 2, 47-51.

[118] ———, The Adams conjecture, Topology 10 (1970) 67-80. MR 43 (1972) 5525.

[119] ———, On the cohomology and K-theory of the general linear groups over a finite field, Annals of Math.(2) 96 (1972) 552-586. MR 47 (1974) 3565.

[120] D. G. Quillen, On the group completion of a simplicial monoid. Privately circulated MS–not to appear.

[121] D. L. Rector, Loop structures on the homotopy type of S^3, in Lecture Notes in Mathematics no.249, Springer 1971, 99-105. MR 49 (1975) 3916.

[122] F. W. Roush, Transfer in generalised cohomology theories, Thesis, Princeton 1971.

[123] B. J. Sanderson, Immersions and embeddings of projective spaces, Proc. London Math. Soc.(3) 14 (1964) 137-153. MR 29 (1965) 2814.

[124] G. B. Segal, Classifying spaces and spectral sequences, Publ. Math. of the I.H.E.S. no.34 (1968) 105-112. MR 38 (1969) 718.

[125] __________, Homotopy-everything H-spaces. Privately-circulated MS.

[126] __________, Configuration-spaces and iterated loop-spaces, Invent. Math. 21 (1973) 213-222. MR 48 (1974) 9710.

[127] __________, Categories and cohomology theories. Topology 13 (1974) 293-312. MR 50 (1975) 5782.

[128] __________, The multiplicative group of classical cohomology, Quarterly Jour. Math. 26 (1975) 289-293. MR 52 (1976) 1667.

[129] J-P. Serre, Homologie singulière des espaces fibrés, Annals of Math.(2) 54 (1951) 425-505. MR 13 (1952) p. 574.

[130] __________, Groupes d'homotopie et classes de groupes abéliens, Annals of Math.(2) 58 (1953) 258-294. MR 15 (1954) p. 548.

[131] R. M. Seymour, Vector bundles invariant under the Adams operations, to appear in the Quarterly Jour. Math.

[132] __________, The infinite loop Adams conjecture, submitted to Inventiones Math.

[133] V. P. Snaith, A stable decomposition of $\Omega^n S^n X$, Jour. London Math. Soc.(2) 7 (1974) 577-583.
MR 49 (1975) 3918.

[134] E. H. Spanier, Duality and S-theory, Bull. Amer. Math. Soc. 62 (1956) 194-203.
MR 19 (1958) p. 51.

[135] ———, Algebraic Topology, McGraw-Hill 1966.
MR 35 (1968) 1007.

[136] E. H. Spanier and J. H. C. Whitehead, A first approximation to homotopy theory, Proc. Nat. Acad. Sci. USA 39 (1953) 655-660.
MR 15 (1954) p. 52.

[137] ———, Duality in homotopy theory, Mathematika 2 (1955) 56-80.
MR 17 (1956) p. 653.

[138] ———, The theory of carriers and S-theory, in "Algebraic geometry and topology", Princeton Univ. Press 1957, 330-360.
MR 18 (1957) p. 919.

[139] J. D. Stasheff, Homotopy associativity of H-spaces. I, Trans. Amer. Math. Soc. 108 (1963) 275-292.
MR 28 (1964) 1623.

[140] ———, A classification theorem for fibre spaces, Topology 2 (1963) 239-246.
MR 27 (1964) 4235.

[141] ———, Infinite loop spaces – an historical survey, in Lecture Notes in Mathematics no.196, Springer 1971, 43-53.
MR 43 (1972) 6911.

[142] N. E. Steenrod, Milgram's classifying space of a topological group, Topology 7 (1968) 349-368.
MR 38 (1969) 1675.

[143] R. Steiner, Smith's Prize essay, Cambridge 1976.

[144] M. Sugawara, A condition that a space is grouplike, Math. Jour. Okayama Univ.7 (1957) 123-149.
MR 20 (1959) 3546.

[145] D. Sullivan, Triangulating homotopy equivalences, mimeographed notes, Warwick 1966.

[146] D. Sullivan, Smoothing homotopy equivalences, mimeographed notes, Warwick 1966/67.

[147] ———, On the Hauptvermutung for manifolds, Bull. Amer. Math. Soc. 73 (1967) 598-600.
MR 35 (1968) 3676.

[148] ———, Geometric Topology, mimeographed notes, Princeton 1967.

[149] ———, Geometric Topology. Part I. Localisation, periodicity and Galois symmetry, mimeographed notes, MIT 1970.

[150] ———, Genetics of homotopy theory and the Adams conjecture. Annals of Math. 100 (1974) 1-79.

[151] R. Thom, Quelques proprietes globales des varietes differentiables, Comment. Math. Helv. 28 (1954) 17-86.
MR 15 (1954) p. 890.

[152] A. Tsuchiya, Homology operations on ring spectrum of H^∞ type and their applications, Jour. Math. Soc. Japan 25 (1973) 277-316.
MR 47 (1974) 4255.

[153] R. Vogt, Boardman's stable homotopy category, mimeographed notes, Aarhus 1970.
MR 43 (1972) 1187.

[154] J. B. Wagoner, Delooping classifying spaces in algebraic K-theory, Topology 11 (1972) 349-370.
MR 50 (1975) 7293.

[155] G. W. Whitehead, Generalised homology theories, Trans. Amer. Math. Soc. 102 (1962) 227-283.
MR 25 (1963) 573.

[156] A. Zabrodsky, Homotopy associativity and finite CW complexes, Topology 9 (1970) 121-128.
MR 42 (1971) 1118.

INDEX

ANNALS OF MATHEMATICS STUDIES

Edited by Wu-chung Hsiang, John Milnor, and Elias M. Stein

90. Infinite Loop Spaces, *by* J. F. ADAMS
89. Invariant Forms on Grassmann Manifolds, *by* WILHELM STOLL
88. Foundational Essays on Topological Manifolds, Smoothings, and Triangulations, *by* ROBION C. KIRBY and LAURENCE C. SIEBENMANN
87. Scattering Theory for Automorphic Functions, *by* PETER D. LAX *and* RALPH S. PHILLIPS
86. 3-Manifolds, *by* JOHN HEMPEL
85. Entire Holomorphic Mappings in One and Several Complex Variables, *by* PHILLIP A. GRIFFITHS
84. Knots, Groups, and 3-Manifolds, *edited by* L. P. NEUWIRTH
83. Automorphic Forms on Adele Groups, *by* STEPHEN S. GELBART
82. Braids, Links, and Mapping Class Groups, *by* JOAN S. BIRMAN
81. The Discrete Series of GL_n over a Finite Field, *by* GEORGE LUSZTIG
80. Smoothings of Piecewise Linear Manifolds, *by* M. W. HIRSCH *and* B. MAZUR
79. Discontinuous Groups and Riemann Surfaces, *edited by* LEON GREENBERG
78. Strong Rigidity of Locally Symmetric Spaces, *by* G. D. MOSTOW
77. Stable and Random Motions in Dynamical Systems: With Special Emphasis on Celestial Mechanics, *by* JÜRGEN MOSER
76. Characteristic Classes, *by* JOHN W. MILNOR *and* JAMES D. STASHEFF
75. The Neumann Problem for the Cauchy-Riemann Complex, *by* G. B. FOLLAND *and* J. J. KOHN
74. Lectures on *p*-Adic *L*-Functions *by* KENKICHI IWASAWA
73. Lie Equations *by* ANTONIO KUMPERA *and* DONALD SPENCER
72. Introduction to Algebraic *K*-Theory, *by* JOHN MILNOR
71. Normal Two-Dimensional Singularities, *by* HENRY B. LAUFER
70. Prospects in Mathematics, *by* F. HIRZEBRUCH, LARS HÖRMANDER, JOHN MILNOR, JEAN-PIERRE SERRE, *and* I. M. SINGER
69. Symposium on Infinite Dimensional Topology, *edited by* R. D. ANDERSON
68. On Group-Theoretic Decision Problems and Their Classification, *by* CHARLES F. MILLER, III
67. Profinite Groups, Arithmetic, and Geometry, *by* STEPHEN S. SHATZ
66. Advances in the Theory of Riemann Surfaces, *edited by* L. V. AHLFORS, L. BERS, H. M. FARKAS, R. C. GUNNING, I. KRA, *and* H. E. RAUCH
65. Lectures on Boundary Theory for Markov Chains, *by* KAI LAI CHUNG
64. The Equidistribution Theory of Holomorphic Curves, *by* HUNG-HSI WU
63. Topics in Harmonic Analysis Related to the Littlewood-Paley Theory, *by* ELIAS M. STEIN
62. Generalized Feynman Amplitudes, *by* E. R. SPEER
61. Singular Points of Complex Hypersurfaces, *by* JOHN MILNOR
56. Knot Groups, *by* L. P. NEUWIRTH
55. Degrees of Unsolvability, *by* G. E. SACKS

Library of Congress Cataloging in Publication Data

Adams, John Frank.
Infinite loop spaces.

(Annals of mathematics studies ; no. 90) (Hermann Weyl lectures)
Bibliography: p.
1. Loop spaces. I. Title. II. Series.
III. Series: Hermann Weyl lectures.
QA612.76.A3 514'.24 78-51152
ISBN 0-691-08207-3
ISBN 0-691-08206-5 pbk.